THE ENVIRONMENTAL DILEMMA

Optimism or Despair?

An Interdisciplinary Analysis of Trends, Issues, Perspectives, and Options

Lambert N. Wenner

University Press of America, Inc.
Lanham • New York • Oxford

Copyright © 1997 by
University Press of America,® Inc.
4720 Boston Way
Lanham, Maryland 20706

12 Hid's Copse Rd.
Cummor Hill, Oxford OX2 9JJ

All rights reserved
Printed in the United States of America
British Library Cataloguing in Publication Information Available

Library of Congress Cataloging-in-Publication Data

Wenner, Lambert N.
The environmental dilemma, optimism or despair? : an
interdisciplinary analysis of trends, issues, perspectives, and options /
Lambert N. Wenner.
p. cm.
Includes bibliographical references and index.
1. Environmental management. 2. Environmental management--
United States. 3. Ecological engineering. I. Title.
√GE300.W46 1997 363.7'05--dc21 97-4496 CIP

ISBN 0-7618-0704-7 (cloth: alk. ppr.)
ISBN 0-7618-0705-5 (pbk: alk. ppr.)

∞™ The paper used in this publication meets the minimum
requirements of American National Standard for information
Sciences—Permanence of Paper for Printed Library Materials,
ANSI Z39.48—1984

Dedication

To my dear wife Inge, for her patience, perseverance, inspiration, and encouragement.

Contents

Preface	xvii
Acknowledments	xxi
Introduction and Overview	xxiii

Part One: Trends Affecting the Environment

Chapter 1: Revolutions in Technology	1
How Technology Stimulates Change Facilitating Population Growth Increasing Economic Efficiency	2
Successive Socioeconomic Systems Hunting and Gathering Horticulture Herding Intensive Agriculture Revolution in Commerce The Industrial Revolution Postindustrial Period	4

Technology and the Environment	**10**

Developed Nations
Developing Nations
Emergence of a Global Marketplace
Two Faces of Technology

Chapter 2: Rapid Population Growth	**13**
The Population Explosion	**13**
The Study of Population	**16**

Demography
Contributions of Other Fields

Current Situation	**18**

Regional Variations
United States and Canada
Other Industrial Nations
Africa
Asia
Latin America

How Credible Are Projections?	**21**
Explaining Rapid Growth	**22**

Reduction of Famine
Suppression of Disease Epidemics
Improved Health and Sanitation
Less Premeditated Killing
Migration to New Areas

Explaining Slow Growth	**26**

Theory of Demographic Transition
Why Fewer Births?

Outlook for Developing Countries	**27**

Chapter 3: Escalating Consumption
 of Resources   29

The Industrial Nations   30
 Leading Consumer Nations
 Goods and Services Consumed

The United States as Role Model   32
 Energy Use
 Automania
 Demand for Goods and Services

Consumption in Developing Countries   36

Industry's Role in Consumption   37

Standard of Living Reconsidered   39
 Redirecting Values
 Quality of Life Redefined

Part Two: Environmental Impacts of the Trends

Chapter 4: Depletion of Natural Resources   41

Renewable and Nonrenewable Resources   42
 Metals and Industrial Minerals
 Fossil Fuels and Gases
 Forest and Forest Products
 Wetlands
 Plants and Animals
 Croplands and Soil Quality
 Fresh Water Sources
 Fisheries
 Unique Natural and Cultural Sites
 Outdoor Recreation

Synthetic Materials	55
Economics of Resource Depletion World Resources Institute	56
Chapter 5: Environmental Degradation	59
An Ecological Perspective	59
Sources of Environmental Impacts Frontier Philosophy of Abundance Industrial Activities Today	61
Key Areas of Concern Mining and Drilling Soil Erosion and Depletion Desertification Deforestation Air Pollution Global Warming Ozone Depletion Water Pollution Ocean Pollution Other Sources of Pollution Species Extinction Solid Waste Disposal Nuclear Wastes	62
Taxing the Earth	77
Chapter 6: Social Effects of **Industrial Development**	79
Global Scale of Economic Growth	79
Charting the Course	80
Costs and Benefits of Development Profit as the Primary Motive	81

Other Benefits of Development
Potential Social and Economic Costs

Economic Growth and Social Change   84
Technologically Induced Change
Urbanization and Its Impact

"Green" Enterprises   86

Persistent Social Issues   87
The Legacy of Colonialism
Status of Women
Ethnocentrism
Antisocial Behavior
Illegal Immigration
Role of Poverty in Population Growth
Environmental Justice
Business Cycles

Chapter 7: Agriculture, Nutrition,
      and Health Issues   97

Global Food and Health Needs   99

Agricultural Trends   99
The Green Revolution
Environmental Effects
The Persistence of Poverty
Limitations on Croplands
Livestock versus Grains

The Paradox of Population Technology   103

Modern Health Hazards   105
Malnutrition
Threat of Epidemic Diseases

Effects of Toxic Chemicals
Global Environmental Change

The Promise of Biotechnology   108
Emerging Applications
Environmental Concerns

Part Three: The Search for Solutions

Chapter 8: The Environmental Movement   111

The Human Role in Nature: Conquer or Coexist?   112

Early Environmentalists   113

The Modern Environmental Movement   115

Sustained Public Awareness   119

The Rise of Extremism   120
Radical Environmentalism
Debunking the Greens

Key Environmental Legislation   121
Wilderness Act of 1964
National Environmental Policy Act of 1969 (NEPA)
Environmental Protection Agency (EPA), 1972
Endangered Species Act of 1973

Major International Initiatives   122

Chapter 9: Optimism or Despair?   125

Claims and Controversy   125

The Continuing Debate   126
Environmentalists

Pro-growth Advocates
Intermediate Views

Prospects for Developing Nations   135

Postscript   136

Chapter 10: The Politics of Reform   139

Barriers to Environmental Action   139
Disagreement among Specialists
Role of Religion and Ideology
Other Traditional Values
National Character
Cultural Inertia

Liberal versus Conservative Policies   143
Conservative or Pro-growth View
Liberal Perspective

Environmental Protection: The U.S. Experience   145
An Auspicious Beginning
The Reagan Reversal
Clinton-Gore Policies
The Republican Congress

Determined Extremists   148
Common Elements of Extremism
Anti-Environmentalists
Radical Environmentalists

Moving Ahead   151

Chapter 11: Bold Actions and Mixed Results   153

Harnessing Human Potential   153

Energy Conservation   154
Wind and Solar Energy

Hydropower
Nuclear Energy
Geothermal and Biomass Sources

Innovations in Transportation 156

Local Initiatives 156
Connecticut River
Humboldt Bay
Sacramento Utilities
Lake Erie

Waste Disposal Quandary 158

Effective Family Planning 159

Results of Environmental Legislation 160
Clean Air Act of 1963 (amended)
Wilderness Act of 1964
National Environmental Policy Act of 1969 (NEPA)
Clean Water Act of 1972 (amended)
Superfund, 1980
Endangered Species Act of 1973 (amended)

International Efforts 164
Montreal Protocol, 1987
Ocean Dumping Ban, 1988
Global Warming Treaty, 1992
Global Fishing Treaty, 1995
World Bank Programs

Dutch National Environmental Plan 166

**Chapter 12: Some Conclusions
and a Look Ahead** 167

The Environmental Dilemma in Review 167
The Triumph of Technology
Environmental Movement

Quest for Knowledge and Understanding 169
Perils of Misinformation
Obstacles to Sound Science and Agreement
Emerging Consensus

Rough Measures of Progress 172
Population Growth Trends
Success of Environmental Programs
International Cooperation
Technological Development

Some Perplexing Issues 175
The Aura of Western Abundance
The Earth's Carrying Capacity
Limiting Population Growth
China's Coercive Population Policy
The Plight of Least-Developed Nations
Costs of Environmental Reform
Costs of a Stable Population
The Need for American Leadership

And What Lies Ahead? 179
The Fanciful
The Feasible

Appendices:

A: Selected Population Data 181
B: Environmental Laws and Agreements 185

Endnotes and Pertinent Readings 191

Index 205

Biographical Sketch 223

Tables and Figures

Table 2-1 Population of the World   19

Table 7-1 Barriers to Social and Economic Well-Being   98

Figure 2-1 World Population Growth Curve   14

Preface

Earth Day 1970 and the environmental movement it spawned inspired countless books and articles about rapid population growth and the plight of the environment. These expressions of concern stimulated a second round of publications from critics who argued that the alleged "crisis" was overdrawn. So why another book?

We live in a time when teeming human populations, sweeping technological changes, and escalating consumption are generating a mind-boggling array of social and environmental impacts. Most experts now agree that many environmental problems are grave and require prompt attention to avoid more serious conditions in the decades ahead. Most of us are fairly well-informed on some of these issues, less familiar with others, and unaware of many more.

Information on these trends and their aftermath is vast, rapidly proliferating, and often contradictory. Few of us have the time to explore much of it. Those who do find that most sources are either unavoidably narrow in focus, too technical for the general reader, clearly biased and self-serving, or inconclusive. There is a continuing need for efforts to summarize, to integrate, and to interpret what is happening so we can put what we already know into a larger context.

Fortunately our expanding knowledge of adverse trends and feasible responses has spurred a wide range of efforts to protect and restore the natural environment. Yet many thoughtful people are concerned that Congress and state legislatures, in their zeal to liberate private enterprise from burdensome environmental regulations, may weaken or even eliminate vital laws and agency procedures that now protect public resources, such as air, water, parks, and forests.

xviii   *The Environmental Dilemma: Optimism or Despair?*

Although these laws and regulations should be reviewed and revised as needed, they have helped to maintain environmental quality. The whole nation loses when commercial development is so unrestricted that short-term financial gains triumph over more critical and enduring public needs.

I am convinced that even more citizens would ardently support reasonable efforts to protect our quality of life if they better understood the problems at hand, were convinced that corrective measures are necessary and feasible, and felt that responsible public and private agencies would act prudently.

This book should help readers to see what they already know in a larger context. It offers a panoramic view of the origins and nature of the population and environment dilemma that:

1. Describes U.S. and world trends in population growth, consumption patterns, natural resource depletion, and resulting environmental stress.

2. Illustrates the complexity of these problems, noting key social, economic, physical, and biological dimensions and how they interrelate.

3. Explains why these conditions and trends are occurring, how different people interpret them, and what is being done to avoid unwanted consequences.

4. Views familiar problems from a broad perspective, so that the reader can decide which are the most significant, which responses seem to be most effective, and what may be in store for future generations.

5. Remains relatively objective, although admittedly concerned, by including varied perspectives of the situation.

6. Is comprehensive, yet concise and informative, with a minimum of technical detail.

Public opinion polls since 1970 consistently demonstrate that we Americans want environmental quality and are willing to pay for it. In particular, we value clean air and water, scenic beauty, and ample opportunities for outdoor recreation. But we are often uncertain on the best course of action to take on environmental issues, as the following example suggests. We use much more fuel per capita than virtually all other nations and now import half of our crude oil. Allowing for inflation, we pay less for gasoline and oil than a generation ago and less than half of what Europeans do. In 1993 the President proposed a national energy tax to conserve fuel and to reduce its environmental impacts. Political opposition was immediate, citing the prospect of rising fuel prices and economic losses from declining petroleum sales, and the measure was defeated.

Many people have since wondered if other considerations were equally important in reaching a decision. A modest energy tax would indeed have increased the price of gas and oil, but potential benefits of the measure were less obvious. The tax could have enhanced public revenues, lowered the annual deficit, and cut imports to reduce our very negative balance of trade with other nations.

Beyond this, such a tax would conserve fuel for future use, encourage more frequent use of existing public transit, reduce air and water pollution (from both transporting and burning fuels), and improve the physical fitness of people who would walk more or use bicycles. Because the tax would save energy by reducing waste and by encouraging the use of more efficient products, future energy costs might be lower than expected. On important issues like this one, all credible aspects should be presented and considered before reaching a decision. Commercial television, radio talk shows, and the popular press sometimes fail to acknowledge the importance of rapid population growth and related environmental consequences. Compare the media coverage of the O.J. Simpson trial in 1995 to the time devoted to all environmental issues that same year.

Serious environmental problems are sometimes unrecognized, deliberately obscured, or mired in controversy because people are ill-informed about conditions and trends, insist on idealistic but unworkable solutions, or misrepresent facts for political gain. Others are afraid (sometimes wrongly) that remedial actions would reduce profits, eliminate jobs, or tarnish reputations. And some individuals fail to acknowledge problems for religious or ideological reasons or because they think that any response is futile.

Few people would deny that we need a healthy environment and adequate resources for the future. This "we" includes giant industries and their investors, local businesses, individual citizens, and the generations yet to come. We have the knowledge and resources to work toward these goals, but often lack the honesty, consensus, or commitment to do what is needed.

Failure to act now can cost our children and grandchildren much more than it would cost us. Our technological potential is impressive, but in the final analysis it is what we accomplish that counts, not what we boast of being able to do.

A final word: An interdisciplinary portrayal of the environmental dilemma is a formidable undertaking and a calculated risk. One trades some specificity and accuracy for breadth and completeness. The overriding goal has been the "big picture," a broad integrated view of major trends and their implications rather than isolated facts and opinions. This involved reviewing historical sources, compiling and interweaving national and global data, identifying key issues, and evaluating current and future outcomes.

The twelve chapters of the book are integrated to provide interdisciplinary coverage of this complex subject. Each is also intended to be complete enough to make sense if read out of sequence, so some redundancy was unavoidable. If the careful reader encounters biases or errors, despite my efforts to avoid them, I hope these are minor and do not detract from the central purposes of this book.

Acknowledgments

Countless individuals have contributed to my background and interest in environmental issues. I am especially indebted to Dave Ketcham, Jeff Sirmon, and Geri Bower in the Washington Office of the Forest Service, Arnie Holden and Jerry Williams in the agency's Pacific Region in Portland, Terry Solberg and the Minerals Management Staff in the agency's Northern Region in Missoula, and Al Young in the Office of Agricultural Biotechnology in Washington for increasing my understanding of agency roles and responsibilities.

I am also grateful to friends and colleagues in academia who have influenced my thinking, especially Hobson Bryan, Bill Freudenburg, Rabel Burdge, and the authors of the many of the publications cited in this volume. I thank Jack Berry for his valuable assistance in preparing the manuscript for publication.

Introduction and Overview

Earth Day 1970 was a dramatic and unprecedented event in American history. Thousands of campus and community activities, many with excellent press coverage, focused the nation's attention on widespread environmental damage and disruption in the wake of relentless industrialization and escalating population growth. Millions of people, including scientists, public officials, nature lovers, journalists, and other concerned citizens, joined forces to dramatize the need to protect the environment.

Earth Day 1990 followed and was international in scale. An estimated 200 million people in the majority of the world's countries participated in parades, ceremonies, teach-ins, and other activities intended to draw attention to the importance of a healthy environment.

Many other nations have taken the offensive to reduce or to reverse destructive environmental trends. In the industrial western nations today, the air is cleaner over many cities, some landscapes are greener, some toxic chemicals are better controlled, and many once-polluted lakes and streams now support fish. Cars get better mileage, newer homes are better insulated, and a significant and growing portion of usable waste is being recycled. The world's population is still increasing each year in record-setting increments but in most nations the percentage of annual growth is beginning to slacken.

The Global Environmental Dilemma

Many well-informed people believe that we are just beginning to stem the swift tide toward environmental calamity. Much remains to be done if more populous future generations are to be assured of a quality environment and sufficient natural resources to meet their daily needs. Unfortunately, scientists, public officials, industry representatives, environmental groups, and other concerned citizens often express profound disagreement on whether problems really exist, the types of action needed if they do, and who should do them. For example, is further population growth a plus or a minus--or both? Do we need massive global action on many fronts or mainly some adjustments in key problem areas? Should government wield a heavy hand to effect change or leave most responses to the private sector? Can industrial nations continue to enjoy their high level of comfort and convenience while developing nations also improve theirs?

Mastering the Environment

For over 99 percent of their four million years on earth, humans were very limited in number and in the scope of their technology. Most societies were very small, often nomadic, and coexisted with other animal species in a rough equilibrium. Humans had to compete with other species and when plants and game were exhausted in one area, they had to move to another. By today's standards, average life expectancy at birth was very short, usually less than a generation, and kept in check by natural hazards, accidents, diseases, intertribal strife, and periodic famine.

Revolution in Technology

Dramatic changes in the conditions of human existence have occurred since the last ice age about 11,000 years ago. At scattered locations around the world, people began to manage their environment by domesticating animals and growing crops to secure their food supply. They learned to utilize a wider variety of plants and minerals for food, shelter, and implements, and developed new forms of social organization to meet emerging needs. This increased their mastery of the environment, permitting larger and more stable societies, as evident in the eastern Mediterranean region, northern India, eastern China, and Central and South America.

About two centuries ago, Europeans learned to harness steam in addition to wind and water power, and the Industrial Revolution was underway. Gasoline and electrical power followed and soon urban-based factories produced goods cheaper and more efficiently than small local shops. By the early 1900s, large-scale manufacturing had displaced agriculture as the dominant economic activity in Europe, North America, and Japan, and stimulated sweeping social and economic changes in these societies.

Introduction   xxv

The development and refinement of industrial and agricultural technology, and related urbanization have enabled far more people to inhabit a given land area. Today over 265 million Americans occupy a land area that supported perhaps a million or two Native Americans and several thousand European colonists three centuries ago. During this same period, the world's population increased about ninefold to 5.7 billion.

Trends That Shape Our Future

This enormous increase in human population has taken its toll on the environment--our biosphere, with its air, water, soil, accessible minerals, and the plants and animals that share it with us. We humans no longer live in harmony with our natural surroundings, but seek to discover, utilize, and control an ever-expanding array of organisms and materials. Our actions both benefit and endanger us and other life forms on this planet.

Growing numbers of scientists, conservationists, and agency officials believe that life on earth now faces more challenges to its survival than at any time since the massive extinction of the dinosaurs and many other species millions of years ago. Three closely-related world trends threaten many forms of plant and animal life, and the equilibrium of the biosphere:

1. Continuing rapid growth of world population, which more than doubled between 1950 and 1996.

2. Increasing per-capita consumption of many nonrenewable resources, including minerals, fossil fuels, primeval forests, fresh water, and fertile topsoil.

3. Cumulative adverse impacts on the social and natural environment from the expanding use of industrial technology.

For decades scholars and conservationists described these trends and predicted their outcomes. But most people were preoccupied with more immediate concerns and, until recently, did not take these early warnings very seriously. Public attention was focused on the Great Depression, World War II, post-war recovery, the Cold War, and conflicts in East Asia and the Middle East. Following World War II, devastated nations had to rebuild their shattered infrastructures, Third World nations attempted to industrialize, and Americans pursued the Good Life by converting production from armament to consumer goods and services.

By the mid-1960s, widespread famine and deterioration of the environment worldwide were too obvious to ignore. Starvation and disease epidemics in the Third World, global air and water pollution, and rising energy prices sharpened our perception of what might happen if present trends went unchecked. Other concerns included intensified cutting and burning of the world's remaining virgin forests, numerous extinctions of plant and animal species, and sharply increased U.S. reliance on imported oil, minerals, and other commodities. Public interest mounted and Congress enacted a series of laws intended to protect the environment. Many states, Canada, and European countries followed suit and ultimately the majority of the world's nations signed various international agreements to reduce environmental deterioration.

Despite some economic arguments to the contrary, most "earth watchers" think there are limits to the amount of natural resource exploitation and degradation the planet's biosphere can endure without serious consequences for humans and other organisms. Although most of these limits are not yet fully understood, a growing body of scientific research indicates that we are at a critical juncture and already have serious environmental problems, some global and others regional or local in scope. A preliminary look at world trends in population, consumption, and technology will help to clarify this.

Growing Population

The world's population approached 5.8 billion in 1996 and the United Nations projects it will reach 8.3 billion in 2025, a generation from now.(1) The 1996 population of the world is twice what it was in 1955, while the U.S. has doubled since 1940. China alone now has more people than the entire world did in 1860, and would have many more without stringent birth control policies. Feeding, housing, employing, and providing basic services for China's 1.2 billion people is already a formidable challenge and stresses their natural environment, even though their total consumption of commercial energy and many other resources is less than in the United States.

Mounting Consumption

Not only are there many more Americans now than a generation or two ago but each one requires more square feet of home, school, and indoor recreation space, more vehicles and miles of streets and highways, more changes of clothes and accessories, more appliances and gadgets, more toys and games, and so on. Each also uses much more commercial energy, buys and wastes more food and expendable products, and disposes of far more containers, often after just one use. These trends increase the per-capita demand for minerals, fuels, forest and farm products, and fresh water.

Some people believe that the continuing increase in per-capita consumption of goods in many industrial and "developing" nations is a more significant threat to the environment than population growth. In both categories of nations, the perceived "need" for personal possessions, housing, infrastructure, comfort, convenience, and related service facilities keeps expanding faster than the population.

Expanding Technology

Modern technology has become very sophisticated with increasingly diverse applications and effects. It can both harm and heal the environment depending on the values of its users. It can destroy lives and property with minimal risk to the user, and do so on a massive scale as during the recent Desert Storm operation in Iraq. It can also save lives on a colossal scale, as when modern antiseptics and vaccines prevent millions of premature deaths from epidemic diseases. Each generation of people lives longer than previously and increases its impacts on the environment when other trends remain unchanged.

Humans have altered the face of the earth with industrial technology by cutting forests, draining wetlands and aquifers, exterminating wildlife, and constructing mines, dams, water reservoirs, highways, canals, buildings, and parking lots. But technology is also used to restore the environment. For example, toxic chemicals are monitored and neutralized, water is tested for impurities and purified, trees are replanted, lakes are restored, air quality is improved, and contaminated soil is reclaimed.

Complexity of the Dilemma

The population-environment dilemma is really a complicated set of problems that resist solutions. We humans possess many different systems of values and often have contradictory goals and perceptions of reality. Our physical and biological environment is very intricate and only partially understood. This complexity is a barrier to understanding the dilemma and, until the last few decades, rather few people grasped its broad scope and seriousness. Widespread scientific and political recognition of many environmental concerns has been slow in coming and support for research in some areas is meager. Even leading scientists sometimes find it difficult to relate their piece of the environmental puzzle to the larger picture. This complexity becomes more apparent when we take a closer look at the global environment.

Major Components of the Environment

Ecologists tell us that the equilibrium of our global ecosystem is seriously threatened as humans increasingly manipulate the environment to satisfy their expanding consumer needs. The global environment includes vast oceans, diverse land areas, thousands of chemical substances, and literally millions of organisms, all subject to natural processes and interacting with each other through time.

The physical component includes air, water (lakes, streams, oceans, and groundwater), soils, terrain, minerals, and fossil fuels. Add to this the planet's natural processes that gradually alter the environment, including prevailing winds and storms, shifting ocean currents, periodic tides, solar radiation, gradual climatic changes, the seasons, fluctuating temperatures and precipitation, continental drift, earthquakes, volcanoes, and objects falling from outer space.

The biological component encompasses over ten million living species, including plants, animals, insects, and microorganisms, each with its own niche in our shared ecosystems. However unique some of these organisms may seem to us, each depends on others for its survival. Each is subject to certain natural conditions, such as the need for air, water, and nutrition, competion for resources, the desire to reproduce, life cycles, and natural disasters. Many species (we are often unsure of which) are essential to human welfare and drastic changes in them would reduce our future options.

The human component consists of billions of people and their numerous social systems, such as families, communities, nations, religions, civic groups, and corporations. We must also include the cultural beliefs, values, goals, laws, technology, needs, and biases that direct and restrict people's behavior in each setting. Humans have physiological needs similar to other species and additional psychological needs. Widely recognized human needs include food, water, shelter, security, belonging, and recognition.

The Critical Human Factor

Global environmental stresses from human population growth, technological alterations, and expanding consumption of resources are now far greater than in any previous generation. Without further remedial actions, many of these impacts will exceed current record levels as populations continue to increase. But humans are also the key to resolving this dilemma because they are capable of evaluating their own behavior and altering their activities if the environmental effects are unacceptable. Most of us anticipate that our descendants will need Planet Earth for thousands of years and that they are entitled to clean air and water, adequate shelter, sufficient food, supplies of minerals and fuels, and various other consumer goods and recreational opportunities. Yet we are mining most of the world's accessible mineral deposits and still discard most mineral products after a few years or decades of use. Each year the world loses millions of acres of prime forest, farmland, and wetlands to logging, strip mining, highways, fire, erosion, commercial recreation, and urbanization.

We also generate mountains of solid wastes and millions of tons of toxic chemicals that almost defy safe disposal. Many are released into our air, water, and soil. These toxic compounds, along with unsustainable hunting, fishing, and forestry practices, have hastened the demise of plant and animal species, perhaps causing the greatest mass extinction since an asteroid assumedly hit the earth with devastating results some 65 million years ago.

Today many governments, responsible industries, and research institutes are trying to learn more about the environmental impacts of human activities and how to avoid or reduce them. Concerned scientists and practitioners increasingly recognize that both social and environmental factors need to be considered if problems are to be understood and workable solutions are to follow. Public agencies and institutes increasingly use interdisciplinary teams to conduct environmental research and to plan and implement viable environmental projects and programs.

Optimism or Pessimism?

Despite impressive gains in our knowledge of the environmental dilemma, one still finds a wide divergence of opinion about its seriousness, the kinds of remedial actions needed, and the prospects for future generations. Scenarios offered by agencies, industries, environmental groups, foundation "think tanks," and individual researchers range from blithely optimistic, to concerned but hopeful, and to deeply pessimistic.

Some Reasons for Optimism

Optimists, many of them religious or political conservatives, argue that we have the technology and resources to meet our present and future needs despite foreseeable population growth. Frequently they cite their trust in God to guide our decisions, human ingenuity to solve our problems, and/or the free market to adapt to new realities. Some say that humans are better off now than they have ever been and that this trend will continue. Environmental problems, when they really exist, can gradually be resolved without increasing government intervention to constrain population growth, to protect or conserve resources, or to change corporate and consumer behavior.

Basis for Pessimism

Many of the specialists who regularly monitor environmental trends fear that some conditions are increasingly threatening and that we are not doing enough to reverse them. Some observers are visibly shaken by the remarkable rate of population growth and the degree to which the world's most accessible natural resources have already been exploited, degraded, and wasted. They report numerous examples of environmental deterioration worldwide and adverse social consequences from them. Unless effective remedial action is taken, they expect a decline in the quality of life of future generations that will have even more people, greater needs, and fewer resources. Some of them foresee drastic resource shortages, severe restrictions to protect a badly deteriorated environment, and bitter conflicts over remaining resources.

An Intermediate View

A third perspective is that the social and environmental impacts of rapid population growth and industrial development are real and pose a growing challenge to scientists, agency administrators, and affected publics. But some threats have been overstated and clearly there is hope for the future if we act promptly and effectively. Numerous public and private environmental programs are already operational in various parts of the world to cope with air and water pollution, waste disposal, land reclamation, wildlife protection, disease control, and other needs. Much more can and should be done to ensure that future human population size and consumption patterns are consistent with the earth's carrying capacity. The free market, left to its own devices, will not do it alone; public and private sectors working together can.

Whatever one's personal philosophy, it is clear that nations in many parts of the world currently lack sufficient technology, accessible natural resources, and effective political and social institutions to provide an acceptable living standard for their citizens. The result is widespread suffering, disillusionment, despair, starvation, disease, and a potential for civil strife and ethnic warfare involving hundreds of millions of people. Each decade we are doing more to resolve these problems but the path ahead still seems long and arduous as more of the globe's peoples sense their deprivation and seek to improve their lot.

Focus of the Book

This book is a broad, interdisciplinary summary of many of the key trends, issues, and options implicit in the environmental dilemma. It is dedicated to students and general readers who wish to broaden their understanding of population trends, environmental consequences, and related controversies.

The first seven chapters describe the causes and consequences of rapid population growth, mounting consumption of natural resources, relevant technological trends, and environmental degradation. Five remaining chapters review and critique various perceptions of the situation, efforts to protect the environment, and scenarios of the planet's future. Technical terms and agency jargon are omitted or explained. Footnotes and additional readings are listed at the end of the book.

No writer can be completely objective, but I have tried to provide a fairly balanced, relatively unbiased view of the environmental dilemma. "Balanced" implies attention to social, economic, physical, and biological dimensions and their interrelationships. "Unbiased" means that I do not speak for any particular interest group and have included a wide range of perspectives of present and future conditions.

This book does acknowledge the seriousness of many current environmental conditions, noting that more than a billion people in various parts of the world lack sufficient resources to meet even minimal standards for food, shelter, and health care. Political reform and economic development could remedy the situation in many cases, but in some countries these solutions are not yet foreseen. In other instances, an unfavorable ratio of people to available resources and technical expertise seems to preclude material standards of living approaching those of industrial nations, at least in the near future.

Yet undue pessimism breeds despair and fatalism, and discourages prompt and effective action to mitigate problems. The outlook for the less-developed nations and their environment is not hopeless because growing numbers of other nations now have the technology and human resources to resolve most major problems and can share it, a trend that is gaining momentum.

Chapter Overview

Part I (Ch. 1-3) reviews trends affecting the environment.

Chapter 1 explores the central role of technological innovation in stimulating population growth, urbanization, consumption, natural resource depletion, and related social change.

Chapter 2 explains how and why the world's population has increased so rapidly in the modern era. It traces a series of cultural adaptations that permit nations to support more people by using efficient technology and a greater portion of available resources. It also introduces demography, the study of population characteristics and trends.

Chapter 3 describes the worldwide diffusion of technology and cultural values, and the increased per-capita consumption of commercial energy, factory goods, and various services.

Part II (Ch. 4-7) surveys environmental impacts of these trends.

Chapter 4 summarizes the growing demand for natural resources and synthetic materials and assesses the extent of national and world resource depletion.

Chapter 5 describes the physical and biological impacts of population growth and technological development, including pollution, global warming, thinning of the ozone layer, species extinction, deforestation, soil erosion, and accumulating wastes. Chapter 6 explains why many people zealously champion worldwide industrialization and others are increasingly critical of it. It describes benefits and costs of development, including jobs, increased business activity, enhanced revenues, and social and economic impacts on local communities and institutions. Chapter 7 looks at present and projected world food needs, efforts to boost agricultural production and nutrition, health hazards and medical services, and future prospects in these areas.

Part III (Ch. 8-12) discusses the search for solutions.

Chapter 8 briefly traces the history of the modern environmental movement in the United States. It reviews contributions of several scientists and conservationists who anticipated the environmental dilemma. It also identifies major initiatives enacted to protect the environment.

Chapter 9 illustrates the great divergence of opinion on the seriousness of population growth and environmental consequences. Failure to achieve consensus thwarts efforts to develop timely, comprehensive, prudent, and flexible environmental policies for the nation and the world.

Chapter 10 identifies some of the obstacles to environmental reform. These include traditional values and value conflicts, self-serving vested interests, insufficient data, political posturing, and unsuccessful past efforts.

Chapter 11 cites examples of recent efforts to protect and to restore the environment in the U.S. and abroad and shows how better research could contribute to more effective policies. It notes that some nations now surpass the United States in their efforts to conserve energy, to recycle materials, and to reduce solid wastes.

Chapter 12 reflects on key points discussed in earlier chapters and addresses some lingering issues. It assesses our progress in resolving the environmental dilemma, including strategies to encourage conservation by government, the corporate sector, and individual families. It also contrasts the fanciful and the feasible as responses to the environmental challenges we face now and in the future.

Both footnotes and pertinent additional readings are listed by chapter at the end of the book. Appendix A contains additional city and country population data. Appendix B provides details on recent international initiatives to protect the environment.

Part One
Trends Affecting the Environment

Chapter 1

Revolutions in Technology

Most people in the United States and other highly industrial nations enjoy a level of comfort and convenience that was unknown two generations ago and is still beyond the reach of most of the world's people. The majority of families have their own house or apartment with central heating, bath and flush toilet, refrigerator, telephone, color television, clothing for each season, and enough income for a for a healthy diet. In the United States, Canada, and several other nations, many families also have two automobiles, a garage, air conditioning, stereo equipment, and personal computers. Although electronic appliances are recent products of our ever-evolving technology, their users increasingly regard them as essential.

This age of plenty in the industrial nations is just a blink in the life span of the human species. Human beings apparently existed in the highlands of east Africa four million years ago and gradually occupied most other parts of the world. People much like ourselves in skeletal structure and brain size, and also using tools and fire, lived in Africa, Asia, and Europe over 100,000 years ago. They spent their often brief lives in nomadic bands or small scattered villages and were prisoners of custom and superstition. They endured extreme weather, famine, and pests with resignation and had little awareness of the rest of the world.

Modern humans (Homo sapiens) inhabited these same continents at least 50,000 years ago and then multiplied, migrated, and sparsely populated the world, including the Americas and various Pacific islands, before and during the last ice age. Despite this wide distribution of humans, it seems likely that more people today live in just one major city, such as

London, Cairo, Mexico City, or Shanghai, than in the entire world until perhaps 10,000 years ago. Since then, human technology has paved the way for an enormous increase in population.

How Technology Stimulates Change

Technology refers to the accumulating fund of knowledge and skills that humans use to meet their needs. Some people extend this definition to include the tools and equipment used to produce goods and services. Humans have created and accumulated knowledge and technical skills throughout their existence, but only in the last one percent of this time have they been able to support millions of people each generation.

For most of human existence, controlled fires, domesticated animals, and human labor were the main sources of energy and many tools, utensils, clothing, and shelter were crude adaptations of materials found in nature. Widely-used materials included leaves, branches, logs, reeds, gourds, stones, shells, hides and fur, bones, sinews, mud, straw, and animal dung. During the past several thousand years a series of technological revolutions occurred that gradually enabled more people to survive by using a larger quantity and variety of the natural resources in a given area, ensuring year-around shelter and food supplies, and reducing deaths due to disease and human violence.

Our ancestors improved living conditions by expanding their technology and developing more efficient forms of social organization. In various times and many places, prehistoric and ancient peoples invented: scraping tools, spears, bows and arrows, clay pots, baskets, robes and blankets, thatched roofs, jewelry, oil lamps, mud bricks, furniture, hearths and ovens, flutes, rafts, sails, wheels, arches, and water wheels. They gradually discovered new uses for available materials, better ways to accomplish tasks, and new forms of energy. With improved productive efficiency, there were more competitive than other tribes and species in their quest for food, shelter, and security, and their numbers increased.

Facilitating Population Growth

Throughout this century, anthropologists have studied hundreds of societies in all parts of the world, varying in size from a few dozen to many millions of people. These societies range from rudimentary to very complex in their agricultural and industrial technologies, their social organization, their cultural heritage, and their ability to exploit the natural environment to meet their needs. Research findings suggest that much of the technology developed by prehistoric peoples to obtain food, shelter, and other necessities is still used in some parts of the world, providing insights about challenges facing earlier generations and their responses to them.

Large human populations were not feasible when all of the earth's people depended on hunting, fishing, and gathering wild plant products for their food supply. Like lions, bears, wolves, and eagles, humans are at the top of the food chain (Ch. 5) and need a large and constant supply of plant and animal products to survive (1). People with scant technology had either to compete with others for locally available supplies, to migrate to less depleted areas, or to die of famine. In many locations they also had to survive cold winters, dry seasons, or other hazards when food was scarce. These persistent needs encouraged humans to develop greater proficiency in language to retain and share what they had learned, and were an incentive for continuing technological development.

Today the worldwide influence of western culture, especially its advanced technology, is pervasive and persistently displacing many traits and techniques of traditional cultures. Modern Eskimos purchase canned goods and clothing at the local store and many who still hunt game use snowmobiles and high-powered rifles.

Increasing Economic Efficiency

"Carrying capacity" refers to the number and variety of humans and/or other forms of life than an area and its resources can support on a long-term basis. Obviously the carrying capacity of a desert differs from that of a fertile valley, but each location will support adapted forms of life. The types and distribution of life forms in each geographic region depend on its climate, terrain, soil, and water sources. For humans, the carrying capacity of an area depends on additional factors, such as:

--the extent of accessible resources and cultural awareness of their potential uses,
--the technology, skills, and capital available for producing foods and materials, and importing goods,
--the quality of human relations, whether in conflict or cooperating, sharing resources, and expanding opportunities,
--the standard of living desired, and ultimately,
--conservation practices that ensure future supplies and a livable environment for succeeding generations.

4   *The Environmental Dilemma: Optimism or Despair?*

As far as we know, humans have developed the most complex language of any species on earth, enabling us to build on experience by trying new methods, to share ideas through verbal communication, and to transmit knowledge from one generation to the next. Thus each successive generation acquires additional knowledge and skills, and has access to a greater variety of plant, animal, and mineral products to feed, clothe, house, and protect its members.

Each new "breakthrough" is from a springboard of earlier inventions and discoveries. Ancient Incas domesticated both animals and plants, such as corn and potatoes. Over the generations they learned how to improve their crop yields by tilling soil, terracing, irrigating, fertilizing, and breeding new varieties. Romans borrowed ideas from the countries they conquered and made further advances in building and road construction, art, literature, and science. Just a century ago, preexisting inventions--the wheel, tire, chassis, clutch, brake, engine, and steering mechanism--were combined to form the automobile. This new invention surpassed each of these earlier components in its impact on modern life (Ch. 6).

Larger societies benefited from having more creative talent and the opportunity to share knowledge and expertise. They developed formal education, improved systems of transportation and communication, and bureaucratic organization for more effective task management through generations of time.

Successive Socioeconomic Systems

Human culture, technology, available natural resources, and population density are closely interrelated, and changes in one affects the others. Even before the last ice age, our resourceful ancestors had begun developing a series of new socioeconomic systems and technological innovations that would enable more people to live in a given land area. Each new revolution in technology also stimulated changes in the social institutions, such as the family, government, economic units, religion, and education in affected societies. These social changes encouraged further technological innovations.

The following illustrations characterize a series of increasingly complex socioeconomic systems that evolved in several parts of the world as human knowledge accumulated, technology became more efficient, and population increased. These descriptions are compiled from numerous case studies to demonstrate the gradual progression from small traditional societies (united chiefly by kinship and largely dependent on local resources) to large modern societies (united by bureaucratic organizations and using resources from all over the world) (2). No doubt some of the reader's ancestors lived within systems similar to each of these.

Hunting and Gathering

The earliest means of subsistence was hunting, fishing, and gathering local plant foods and materials. These activities require few people, little task specialization, and limited technology, and seem to have been the dominant pattern worldwide for most of human existence. Some favored locations, such as the northwest coast of the United States, the Mediterranean, and coastal China, provided an abundance of plant and animal foods throughout the year and permanent settlements were feasible. But in much of the world, hundreds of acres were needed to support a single family with these methods, explaining why societies tended to be small (bands or tribes), nomadic, and only loosely united with others in the area.

Some groups traveled long distances when resources were depleted or when herds of game animals migrated. En route they lived in crude temporary shelters, although some spent part of the year in more permanent lodgings near other bands. Recent examples of hunting and gathering societies include the Plains Indians (following buffalo herds), Eskimos of northern Canada and Alaska, aboriginal peoples of Australia, and various tribes of the Amazon and Congo basins.

Horticulture

About 10 to 20 thousand years ago, people in several parts of the world learned to augment hunting and gathering by planting crude gardens with locally available plants. Despite little or no preparation of the soil, plants such as corn, beans, yams, squash, and cassava were grown in spots selected for adequate moisture and, with luck, crops were harvested. People also learned to dry, salt, or bury foods to preserve them for later consumption, thus reducing the need to migrate in search of new sources of plant foods or game.

Some horticultural peoples gradually improved crop yields by tilling, watering, and fertilizing them, and were then able to remain in one place for generations. This encouraged construction of permanent villages with durable homes and more sophisticated equipment and techniques, such as pottery making, basket weaving, ovens for baking bread, and canoes for river travel. With horticulture to supplement hunting and fishing, a given region could support more people and the population grew. The Pueblo Indians of the southwestern United States and Mexico, some early Mediterranean peoples, and many Polynesians in the South Pacific depended primarily on horticulture for their subsistence.

Herding

During this same period, people began to domesticate dogs, cats, pigs, and grazing animals, eventually including cattle, camels, horses, goats, sheep, camels, yaks, llamas, and reindeer. By protecting these animals from predators and helping them to survive winters or dry seasons, they had a dependable supply of meat, milk, eggs, skins, bone, and other materials, and could also use beasts to pull or carry loads and for security. This was especially important in the colder and drier parts of the world where other forms of agriculture were impractical. Herding societies were often nomadic due to the forage needs of their animals. Many also hunted wild game and gathered plant foods to supplement their diets. Lapps, Bedouins, traditional Basques, and the Masai in Africa exemplify this lifestyle even today.

Intensive Agriculture

Beginning perhaps 10,000 years ago in Egypt, south and east Asia, and later in the American tropics, humans learned to breed and cultivate plants, including rice, wheat, corn, millet, and potatoes, to produce higher-yield crops. They also developed crude implements, such as hoes, rakes, and plows to prepare and fertilize the soil. Later they learn to terrace slopes to capture rainfall and reduce erosion, and to irrigate their fields by diverting stream water into canals.

Ultimately intensive agriculture became efficient enough for families to produce food surpluses for barter or sale, and some people were able to enter other occupations, such as merchant, craftsman, soldier, priest, or government. Permanent towns and small cities were more feasible, along with larger units of political and religious organization. This type of economic system dominated Europe and North America until a century or two ago and is still a major source of employment and subsistence in much of south and east Asia, Latin America, Africa, and other parts of the world.

Revolution in Commerce

During much of human existence, the exchange of goods and services was severely limited by geographic barriers, inefficient transportation, and cultural differences in language, tradition, and attitudes toward strangers. Local bartering increased the range of consumer choices, even though face-to-face contact with other societies was sometimes avoided. As societies gradually enlarged through conquest, common interests, and intermarriage, potential markets expanded and trade increased.

With the emergence of empires, beginning around 3000 BC in ancient China, India, the eastern Mediterranean, and tropical America, people gained access to a much larger variety of natural resources and products. Laws to ensure public safety were promulgated and enforced, and commerce flourished. This encouraged public markets, road and trail construction, the development of trading companies, animal caravans and ships to transport goods, written languages and numerals, barge canals, improved roads, and wheeled vehicles.

Eventually, as the merchant class grew in size and influence, money was coined, task specialization increased, banks were established, and credit was available for business expansion. All of these developments paved the way for the Industrial Revolution that followed.

The Industrial Revolution

Beginning over two centuries ago, the pace of invention and discovery quickened as western nations shifted from a reliance on human and animal muscle, augmented by wind and water power, to machines powered by fuels. This Industrial Revolution first occurred in England, where all of the essential prerequisites were available, including a stable government, a tradition of free enterprise, large domestic markets, skilled craftsmen, a literate class, roads and waterways, a merchant fleet, a navy to protect trade routes, and abundant sources of energy and minerals.

A generation later this revolution spread to the United States, Germany, France, Benelux nations, Canada, and Italy, countries with similar economic potential. Increased mechanization and the use of new materials and manufacturing processes resulted in many new products, lower prices, and an increased demand for natural resources from other countries. By 1900, industrial technology was widely used in most of Europe, Russia, Japan, South Africa, Australia, and southern South America.

Powerful new industries emerged, pursuing specific corporate goals and exerting a growing influence on national and international affairs. The quest for new markets for factory goods and additional sources of raw materials intensified trade, colonization, and international operations. In the transition, productive efficiency was greatly increased and many of the most dangerous and back-breaking jobs in mining, manufacturing, construction, and agriculture were eventually eliminated.

The revolution in energy sources hastened industrial development. With the development of the steam engine, the railroads, the stoker furnace, and electrical power in the 1800s, coal quickly surpassed wood and water power as the leading source of energy. Like wood, coal is a messy fuel, and produces large amounts of soot, ash, clinkers, and noxious gases. Since World War II, oil and natural gas have become the fuels of choice in the industrial nations, because they are clean, cheap, easy to transport, and presently abundant. Hydropower is also important in eastern Canada, the Pacific Northwest, central Africa, and other areas where flowing water is abundant.

Nuclear power generation began in the early 1950s and power plants are now found throughout the United States, Europe, and Japan, and also in Russia, south and east Asia, Canada, South Africa, Argentina, and Mexico. However, public concerns about safety and thermal pollution (heating of surface waters) are now limiting its expansion. Environmentally friendly solar and wind power generation are increasing in popularity but still account for a small fraction of the total energy consumed. New forms of energy, such as hydrogen engines and nuclear fission, are not yet commercially feasible (3).

The Industrial Revolution soon gave birth to an agricultural revolution (Ch. 6-7) as tractors and machines replaced horses, oxen, and manual workers on many farms. Displaced people moved to the cities and within two generations, from about 1900 to 1960, rural residents of the United States declined from 60 to 30 percent of the total population. Today 25 percent of Americans are officially classed as rural but most live in small communities or along highways and only two percent actually reside on farms (4). Similar, if less drastic, rural to urban migrations occurred in other industrial nations and this trend is now very evident in most developing nations as well.

Postindustrial Period

This is the emerging pattern of social and economic life in long-established industrial societies, such as the United States, Canada, and the European Community. Industrial efficiency is now so great that only a fourth of the work force is needed to produce farm, factory, mineral, and forest products. Well-paid factory and middle management employment is being scaled back and lower-paid service jobs are increasing in number. Most employees now provide public and commercial services, such as wholesale and retail sales, transportation, public safety, health care, data processing, banking and investments, insurance, education, social services, maintenance activities, and various personal services (e.g., hair styling, pizza delivery, and dog sitting).

A key development of this period is the electronics revolution, especially the advent of computers, and the resulting changes in manufacturing, transportation, communication, data processing, and entertainment. The 1948 invention of the transistor for sound amplification and switching functions made fragile, cumbersome vacuum tubes obsolete and opened the door to a new generation of sturdy, compact, and inexpensive solid-state electronic appliances and equipment. Some applications:

--Automation: the use of computers, robots, and other devices to control machines and other equipment used in the manufacturing process. Automation displaces many semi-skilled production line workers but also creates (sometimes fewer) new jobs in manufacturing, selling, and maintaining electronic equipment. Depending on the nature of the work, one person at a modern factory or major construction site can replace 2-20 workers in a more conventional work place.

--Data processing: the use of computers for data collection, word processing (writing), data storage on magnetic discs, sorting and classifying data, calculation, statistical evaluation, projection (e.g., predicting weather, economic trends, or the population in 2020), and data transmission via telephone lines (e-mail).

--Communications: the use of compact disc recorder-players for vocal and musical productions, videotape camera-recorders for films, and portable cellular phones for personal communication. In addition, cable television increases the number and clarity of channels, and earth satellites relay television broadcasts to distant locations and track world weather patterns.

--Transportation: the use of computers to guide missiles and space vehicles, to place satellites, to coordinate airline schedules and operations, to increase the efficiency of engines and diagnose problems, and to guide modern subways and commuter trains.

These several descriptions of past and present socioeconomic systems oversimplify reality. The social and economic history of each society is unique because of differences in population size, ethnic diversity, extent of trade, conquest by others, religious beliefs and cultural values, migration patterns, missionary activity, and other factors. In modern times, change is also instigated by newer institutions, such as strong central governments, large corporations, the mass media, commercial advertising, scientific research, and formal education.

Technology and the Environment

Developed Nations

The term "developed" or "industrial" nation usually denotes a country with a well-developed industrial sector, a literate, technically skilled work force, high standards of health and sanitation, and a modern infrastructure (for example: homes and workplaces with electricity, plumbing, and adequate heating and lighting, paved streets and highways, and upscale schools, hospitals, shopping facilities, and services). Long-standing industrial nations, such as the United States, Canada, Japan, and the countries of western Europe lead the world in their production and consumption of commercial goods and services. After World War II, another dozen nations, scattered from eastern Europe to southeast Asia, rapidly expanded their industrial sectors and now produce and export large quantities of factory goods. A few now rival older industrial nations in their per-capita production of goods and services and easily exceed most other developing nations (Ch. 3).

Developing Nations

Developing nations lag behind the industrial West in per-capita industrial production and earnings but are attempting to expand and modernize their economies and to improve literacy, health, housing, utilities, and other social and material conditions of their citizens. Most of the world's 190 or so sovereign countries are now exploiting their minerals and forests, expanding their industrial sectors, and upgrading their infrastructures to compete with the industrial nations (5). Many currently export a large portion of their vital raw materials, such as timber, minerals, and petroleum, at bargain prices to obtain the capital to modernize.

The challenge to modernize is formidable and in some developing nations progress is slow in coming. In countries with rapid population growth, children easily outnumber adult wage earners and taxpayers, unemployment tends to be high, wages are low, and public revenues are meager. Some of these nations are former colonial territories in Africa, Asia, and Latin America that have recently gained their independence and lack financial resources and technical expertise. But they recognize their needs and seek foreign loans, investments, and technical aid to achieve their goals.

In many of these countries, industrial development, environmental protection, and social reforms are impeded by widespread poverty, illiteracy, religious extremism, political corruption, and/or restricted freedom (Ch. 6, 7). Oil-rich countries have a distinct advantage when they have the foresight to invest their enormous royalties prudently.

Emergence of a Global Marketplace

Numerous multinational corporations based in Canada, North America, Europe, and Japan have extended their mining, manufacturing, and marketing operations around the globe. "Exporting" manufacturing abroad is profitable because industrial employees in developing nations usually work for very low wages, foreign markets are expanding, laws and policies restricting pollution and resource exploitation are sometimes much less stringent, and critical raw materials may be cheaper. Foreign operations augment private and nationalized enterprises in the host countries, hasten industrial development, increase their job opportunities, and may also generate adverse social and environmental impacts (Ch. 5-6).

Prior to the 1950s, most large countries, such as the United States, the Soviet Union, China, and Germany, were fairly self-sufficient, producing most of their own consumer goods, exporting some surplus production, and limiting imports mainly to raw materials and products not domestically available. This situation is rapidly changing. In 1950, for example, the U.S. imported only $8.9 billion in goods and exported $10 billion. Between 1950 and 1994, however, the current dollar value of U.S. imports increased over seventy times while exports rose fiftyfold, reflecting not only inflation and and a weakened dollar, but much greater foreign trade and far more dependence on other countries' exports (Ch. 3). World export trade increased from $314 billion in 1970 to $3630 billion in 1993, 11.5 times greater (in current dollars), while population increased only 50 percent (6).

More and more manufactured products are the result of integrated efforts in several countries. An automobile may be assembled in one country from parts supplied by several others. Clothing may be designed in the U.S., made of cloth woven in Taiwan from Australian wool or German synthetic fibers, assembled in Bangladesh or Sri Lanka, and sold in Europe. Increasingly each country in the world produces what it does most efficiently (or cheaply) and imports most of the rest. This trend, evident for decades in the expanding common market economy of the European Community, is now global, bolstered by new trade agreements.

In 1994 the North American Free Trade Agreement (NAFTA) reduced trade barriers among Canada, the United States and Mexico. In 1995 over 100 countries agreed to liberalize their trade policies under the revised General Agreement on Tariffs and Trade (GATT) administered by the World Trade Organization (WTO) (7). Traditional international obstacles to trade such as tariffs, wars, piracy, inconsistent production and measurement standards, and the short life of perishable commodities are gradually being eliminated.

Two Faces of Technology

Critics of unrestricted development persistently argue that the cumulative adverse environmental effects of industrial technology are becoming very serious (Ch. 4-5, 8-9). Despite the tremendous potential of technology to do "good" by meeting human needs, it is most readily applied when it will earn a profit for investors. Some industries have learned to profit from protecting the environment even though many other firms still resent such requirements and try to influence legislators to weaken them (Ch. 6, 10). Technological options include expanding nonpolluting energy sources (e.g., solar, tidal, and wind), producing equipment for improving air and water quality, improved logging and reforestation techniques, and recycling. Government agencies develop environmental protection procedures and mandate their use in the corporate sector to get desired results (Ch. 8). Some firms see the need for these regulations and simply comply. Others try to circumvent them or complain that they are excessive or too expensive to implement, while taxpayers are reluctant to foot the bill. Although such measures may increase manufacturing costs and reduce short-term profits, they also ensure more adequate future supplies of raw materials for both industries and the nation.

Chapter 2

Rapid Population Growth

The Population Explosion

During the past 150 years, the earth's population has increased remarkably fast, from about 1.1 billion people in 1840 to 5.8 billion in 1996. This is a fivefold increase within six generations, half it occurring since 1955. Recently the rate of growth has decreased, from 2 percent annually in the 1960s to about 1.7 percent in the 1990s, but the total number of people added each year continues to increase. World Resources Institute reported an average annual gain of 86 million between 1990 and 1995, and the Population Institute estimated that world population increased almost 100 million in 1995. This one-year gain is more than the population of Germany, Denmark, and Sweden combined.

Most of the decline in the world growth rate is due to the successful family planning programs of a few countries, such as China, Thailand, Taiwan, and Indonesia. If family planning programs in 80 other countries remain relatively ineffective, they could double their populations in 30 years or less and increase total world population to 11-13 billion before tapering off (1). If they are more effective, world population could stabilize at 8-10 billion by 2050.

Worldwide urbanization has rapidly intensified since 1900, when only about 10 percent of the planet's inhabitants lived in metropolitan cities. Urban residence increased to about 30 percent by 1950 and was close to 50 percent in 1996. Cities are getting much larger as well as more numerous.

14   *The Environmental Dilemma: Optimism or Despair?*

Figure 2-1

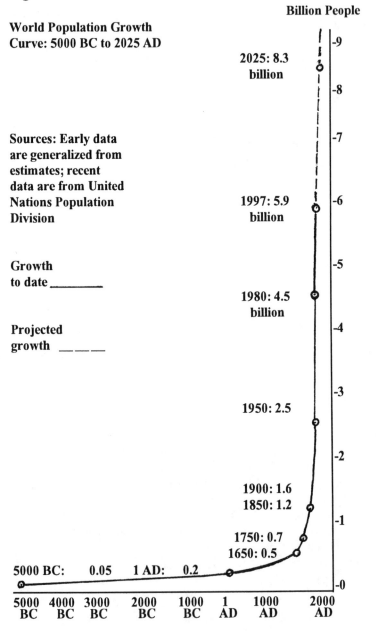

The number with over 5 million people increased sharply from 10 in 1950 to 33 in 1990, and 15 of them exceeded 10 million. Five of these 33 cities are in the U.S., 4 are in Europe (including former U.S.S.R. nations), and 22 are in developing nations (Appendix A).

Figure 2-1 depicts estimated world population at different dates during the past 9000 years. Compare the very gradual increase until about 1800, even though families were generally large, with the soaring growth of the 1900s. Then note the projected increase to almost double the present population within the next 50 years and possible increases beyond that date. People are often surprised when they view this curve of recent and projected population growth. Prior to exhibiting this chart, the author asked a sociology class to estimate what year the world was half as populous as today. Most students thought it was before 1900 and some ventured guesses as far back as the Roman era. After seeing that the date was (then) 1950, some students asked why such rapid growth occurred when effective methods of birth control were increasingly available.

The consequences of rapid growth are critical because each doubling of the world's population sharply increases impacts on the natural environment. World consumption of fossil fuels (coal, gasoline, oil, etc.) and other natural resources has been rising for decades and resulting environmental degradation is apparent worldwide. The earth's natural processes (e.g., sunlight, rainfall, decay, plant activity, and soil regeneration) usually can "repair" damage to the environment from a few hundred million people, as in the past, but constantly increasing the total population stresses ecological systems, hampers their ability to recuperate, and forces humans to take additional remedial actions to protect environmental quality (Ch. 4-5).

If we are to meet the awesome challenges posed by massive population growth, we need to understand the extent of this growth, what causes it, its probable consequences, and what can be done about it. We must also determine the earth's "carrying capacity," the number of people it can support now and in the generations ahead. This capacity depends on the material living standard we expect to maintain, the state of our knowledge and technology, and our willingness to do what is needed. Obviously the world could support far more people at India's current level of consumption than at the current U.S. standard. India, with over three times the population and a third of the land area of the United States, has ten times the population density but possesses and consumes smaller quantities of many vital resources.

The Study of Population

The reader may wonder, how accurate are these estimates and projections of population growth? A review of some previous work will help to clarify this and help to introduce several key concepts and assumptions used in population analysis.

Back in 1798, Thomas Malthus, an English economist and social theorist, wrote that human population tends to increase at a geometric rate, but it is kept in check by war, famine, disease, and (added later) moral restraint. He foresaw a potential for more rapid population growth as humans learn to conquer these limitations and also relieve population pressure by emigrating to sparsely populated lands, such as the Americas and Australia. But Malthus believed that eventually population would outrun the available food supply (2). In the years since Malthus, various critics have discredited portions of his theories, but his insights were remarkable for that time. His systematic analysis set an example for future social scientists.

Fortunately our knowledge of the "population explosion" and its effects is rapidly expanding. This broad subject piques the interest of people in many scientific and applied fields, including agriculture, health, economics, sociology, forestry, engineering, medicine, and zoology. Some of these disciplines collect and analyze population and natural resource data for general reference, while others use it in their work.

Demography

The core field of population study is demography, the statistical analysis of population characteristics. Demographers calculate the size, composition, and distribution, of past, current, and future populations using a variety of sources, including census and other survey data, official records of births, deaths, migration, and other sources. They also assess the causes and effects of population trends and provide useful concepts, such as the following, for discussing these changes:

--Crude birth rate, the number of children born each year per 1000 people, a common measure of a country's fertility. --Fertility rate, the average number of children that women in some defined group have in their lifetime.

--Crude death rate, the number of people per 1000 who die each year, a measure of mortality, or deaths.

--Doubling time, the number of years needed for a population to double in size, especially informative because it reflects both birth and death rates, and usually migration as well.
--Median age, the age that separates the older half from the younger half of the population.
--Migration, relatively permanent relocation in a different country or region.
--Immigration, migration into another country.
--Emigration, migration out of a country.
--In-migration and out-migration, movement within a country or region, such as rural to urban, or city to suburb.

Historically age, sex, and marital status have been important indicators of population growth potential. Societies that encourage all women to marry and at a young age will tend to have higher birth rates than other societies. The median age of a population is a clue to its birth and death rates and hence its potential for growth. A median age of 32 or higher suggests a combination of very low death rates, low birth rates, and a large population of elderly people. A median age as low as 18 suggests a high birth rate, a relatively short life expectancy due to above-average death rates, and a potential for rapid growth as the children mature. This growth could encourage urbanization and increase the demand for schools, public services, and jobs for the additional people.

Just three trends, fertility, mortality, and migration account for population trends in a country or city, and these variables are used to project changes. Nations with very high birth rates and low to moderate death rates may double their population in less than 25 years. Kenya, Libya, Zambia, Iran, Kuwait, Pakistan, and Paraguay are examples. Some other countries have birth rates so low that, despite very low death rates, their populations do not increase at all. Belgium, Germany, Denmark, Austria, Bulgaria, and Hungary have virtually zero population growth (3).

Contributions of Other Fields

The current demand for population data goes far beyond the quest for numbers and trends. Many fields of study augment demographic findings by explaining how and why various trends occur and by interpreting the significance of these changes for social institutions, such as the family, education, health, government, the economic sector, recreation, law enforcement, and religion.

The "baby boom," the sharp increase in the U.S. birth rate in the decade following World War II, illustrates the importance of interdisciplinary findings. At the time, demographers told us how many more babies were born and sociologists explained why. For example, the economy was strong, there were more parents than usual (because of both younger marriages and returning veterans), and people were moving to larger homes in the suburbs for an improved family life. Economists noted how the baby boom affected the type and volume of retail sales and expanded the need for schools and other youth services. Sociologists cautioned us to expect increasing autonomy for young people and more delinquency. Political scientists observed the voting characteristics of the maturing baby boom generation and its effects on elections and the functions of government.

Natural science fields, both theoretical and applied, influence population changes and help us to adapt to them when they occur. Medical researchers save lives, thereby increasing life expectancy, and also develop effective techniques for birth control. Geologists, foresters, and agronomists try to ensure adequate future supplies of minerals, forest products, and foods. Hydrologists are concerned about groundwater pollution and falling water tables (Ch. 5). Each decade more disciplines become involved in population and environmental quality issues.

Current Situation

Even though the rate (percentage) of population growth is now beginning to decline due to more effective family planning in some countries, total population continues to increase because so many people now survive childhood and become parents. Over 90 percent of anticipated growth is expected in developing nations, where three billion young people are or soon will be in their prime reproductive years (4). Globally, the great majority of people are either children (31.5 percent under 15) or young adults, so family size reduction is critical to curbing the future population boom (Ch. 7). Scientists and public officials differ widely in the significance they attach to this potential for growth (Ch. 9).

Regional Variations

Growth rates differ widely among nations, roughly varying with the degree and duration of industrialization. The population in almost all older industrial nations now gains less than one percent annually from natural births and several have stable populations. Newer industrial nations tend to have somewhat higher birth rates and developing nations have the highest, some doubling their populations in 25 years or less.

Table 1 summarizes growth trends by major world region. Appendix A provides more details for selected countries, including past and present populations, the annual rate of increase, and the projected increase.

Table 1

World Population (estimated in millions)

Region	1750	1850	1950	1995	Annual % Chg	2025
North America	5	39	220	454	1.4	616
South America	7	20	112	320	1.7	463
Europe	140	265	549	727	0.2	718
Asia	476	754	1403	3458	1.6	4960
Africa	95	95	224	728	2.8	1496
Australia, Oceania	2	2	13	29	1.5	41
World	725	1175	2520	5716	1.6	8294

Sources: U.S. Census Bureau (for 1750 to 1850) and United Nations Population Division (for 1950 to 2025)

United States and Canada

The population of the present area of the United States grew rapidly from roughly 5 million at the time of the first census in 1790 to about 267 million in 1996. Half of this increase has occurred since 1940 and the Census Bureau projects a further increase of 44 percent (middle series estimate) in another 50 years. Canada, with a population of 29 million, is growing 1.1 percent annually, a bit faster than the U.S. at 1.0 percent. Immigration contributes significantly to the growth of both countries, and many newcomers follow the large family tradition of their native country for at least one generation.

Other Industrial Nations

Europe currently has the world's most stable populations. Most of its 36 countries gained or lost less than 0.5 percent annually from 1990-1995, compared to the current world average of 1.6 percent growth each year. Voluntary family planning has been effective in all social classes and in most religious and ethnic groups, so the traditional large family is unusual.

Predominately Roman Catholic countries, including Spain, France, Belgium, Italy, and Austria, now increase at about the same modest rate as neighboring Protestant countries. Italian families average 1.3 children, among the smallest on earth. The growth rate in Japan is comparable to western Europe and most rates in eastern Europe from Poland to Greece are even lower.

Growth rates in other industrial nations, including Australia, New Zealand, South Korea, and Singapore are generally higher than in Europe but below the world average. Industrial nations are now a sharply declining segment of world population and are likely to decrease from 31 to 14 percent of the total during 1960-2025, a span of about two and a half generations (5).

Africa

Africa's growth rate increased substantially during the past generation from an annual average of about 2.6 percent in 1965-1970 to 2.8 in 1990-1995. If this growth rate continues, Africa's population is projected to increase from 281 million in 1960 to 1500 million in 2025, over five times greater in 65 years. Growth rates for 37 of Africa's 48 countries have increased since 1960. Angola, Cote d'Ivoire, Libya, Malawi, Uganda, Zaire, and many others are growing very rapidly and only a few nations are reducing their growth rates.

Asia

Most Asian nations are growing at rates of 1.5 to 4.0 percent annually. The highest rates are in the Mideast, where some countries have oil revenues to supplement otherwise modest economies. During the past few decades a few countries, including China, Singapore, South Korea, Sri Lanka, Thailand, and Taiwan, have sharply reduced their growth rates to below the world's average.

Most of the recent reduction in the world's population growth rate can be attributed to China's stern family planning program with penalties for producing more than one or two children. China had 21 percent of the world's people in 1990 and is projected to have only 18 percent by 2025.

Latin America

Population growth in Mexico and most countries of Central and South America exceeds the world average, although Argentina, Uruguay, Cuba, and several other Caribbean nations have below average growth. Generally high birth rates in the past generation and a sharp decline in infant deaths resulted in a large number of young women, so the potential for additional growth in most of these countries is high.

How Credible Are Projections?

Valid short-term population projections are extremely important to state and national governments, county and city planners, school officials, pension programs, the insurance industry, manufacturers, retail businesses, and others who must plan to accommodate a changing population.

Critics of population projections claim that future estimates are unreliable because current data are sometimes inaccurate and people change their behavior over time. Statisticians try to minimize these problems by verifying census counts with spot checks and by offering two or more sets of projections, based on different assumptions. Admittedly, the longer the term of the projection, the greater the risk of errors. But carefully formulated short-term projections usually are reasonably accurate. Changes in birth and death rates tend to be gradual in most societies and these trends, along with other developments that may affect population changes, are considered before making projections.

In estimating the population of entire continents or the world, small errors in national data often neutralize each other. When world population passed the 3.2 billion mark in 1963, the United Nations projected an increase to 6.13 billion by the year 2000. The revised projection twenty-five years later was 6.25 billion, just two percent more. The still more recent 1994 projection for 2000 by the U.S. Bureau of the Census is 6.17 billion, remarkably close to the 1963 U.N. estimate. Most estimates for individual nations were fairly consistent with later findings but some estimates had to be revised upward and others downward when birth rates changed more than anticipated.

Explaining Rapid Growth

Until the early 1900s, a high birth rate was the norm almost everywhere, even in Europe, but it was offset by a medium to high death rate which limited net growth. Large families were the cultural ideal and are still common in more traditional societies and ethnic groups. Even where smaller families prevail, parents are routinely congratulated after each new birth, babies are roundly admired, and some men regard numerous children as proof of their masculinity, even when they fail to contribute to their training and support.

Prior to the 1950s, most of the world's parents were rural and usually regarded their offspring as an economic asset and as a source of support when they were retired. Children helped to gather food, maintain gardens, herd livestock, till fields, stack hay, harvest crops, gather firewood, and do household chores. Most children did not attend high school or college, nor did they acquire expensive tastes for clothes, cars, appliances, and recordings, and were not considered a serious financial burden. If the net population gain in each generation were modest and land was available, additional people were accommodated by expanding acreage, increasing yields, getting assistance from other family members, or emigrating to more promising areas.

These traditional options diminished as rural populations grew more rapidly, suitable land became more limited, and machines displaced farm workers. Europe and North America suffered gravely from surplus farm and factory production, rural poverty, and joblessness during the Great Depression of the 1930s. Today rural areas in many parts of the world can no longer support additional people in the traditional ways and millions of young adults flock to the cities and to an uncertain future (Ch. 6-7).

The population "explosion," the 500 percent increase in world population during the past 150 years is due mainly to the sharp worldwide decline in the death rate. Historically, famine, disease epidemics, and insanitary conditions took a heavy toll, especially among infants. This high death rate was drastically reduced by a series of social and technological innovations (especially between 1870 and 1950) that doubled the life expectancy of newborn children in industrial countries.

Since 1950, further advances in education, transportation, communication, medicine, and marketing, and the efforts of numerous international aid programs have helped to increase life expectancy worldwide. Major developments that reduced the death rate include:

Reduction of Famine

Many scientists believe that famine was once the leading factor in population control. Even though many societies still experience widespread starvation and malnutrition, a declining percentage of the world's people die from food shortages. The continuing development of high-yield plants, application of fertilizers to exhausted soils, improvements in food preservation methods, expanded marketing and delivery services, and public and private aid programs help to ensure that more people survive (Ch. 7).

Suppression of Disease Epidemics

Historians cite numerous examples of disease epidemics all over the world dating back to ancient Egypt and other early civilizations. The most severe epidemic on record, the notorious Black Death (probably bubonic plague), raged through east Asia and Europe about 1330 to 1350, killed an estimated half of the population, and kept returning periodically to take more victims. During the period of European colonization in the Americas, Africa, and Polynesia, millions of native peoples died of common diseases, such as measles and smallpox, against which they had little innate immunity. In more recent times (1918) an especially virulent form of influenza killed 20 million people. Many highly contagious and potentially fatal diseases, such as cholera, bubonic plague, typhoid fever, polio, smallpox, malaria, and yellow fever, have been contained. This was accomplished through the development and use of antiseptics, vaccines, pesticides, and various medical procedures to reduce contagion. Scattered outbreaks are quickly suppressed and when new diseases such as AIDs or ebola emerge, medical researchers seek ways to control them (Ch. 7).

Improved Health and Sanitation

In past centuries, over half of the infants in many societies failed to survive the first few years of life. Adults also died from conditions that can easily be remedied today, such as organ failure, childbirth, serious injuries, or infections. Each decade brings advances in diagnosing ailments, medicines, therapies, surgical procedures, and health care facilities. Hospitals, medical clinics, or first aid stations now exist in virtually all cities and many smaller communities around the world. A growing portion of people live in urban areas and have access to medical care, while improvements in public transportation and expanded social programs increase the options of rural people and the poor.

Other technical advances that increase longevity include water purification and distribution, sewage and solid waste disposal, and food processing. Laws have been enacted to require quarantines for highly contagious diseases, to improve quality control in food and drug manufacturing, to set and enforce health and nutrition standards, and to protect environmental quality (Ch.7-8). Compared to previous generations, a far greater percentage of children now survive and become parents themselves.

Less Premeditated Killing

In most times and places, premeditated killing, legal and illegal, was a significant check on population growth. Examples include wars, genocide, civil strife, feuds, duels, pogroms, murders, burning or burying widows and servants, killing people judged to be defective, and executing criminals, heretics, witches, and political dissenters.

Some famous conquerors, including Genghis Khan, decimated both military and civilian populations that resisted their offensives. In medieval Europe, hundreds of thousands of people were condemned to death for alleged heresy and witchcraft. The Thirty Years War devastated Germany and a large portion of its population was killed. Some societies permitted private citizens to kill others for reasons such as adultery, rape, robbery, or insubordination. Criminal executions were also much more common than now; early in the 19th century over 200 English crimes carried the death penalty.

More recent examples of large-scale extermination of people are Stalin's slaughter and starvation of several million peasants and political nonconformists in the 1930s and 1940s, the death of an estimated 50 million people during World War II, and the execution or starvation of perhaps another 50 million during the Chinese civil war, the Japanese occupation, and later when Mao Zedong (Mao Tse-tung) emerged victorious and restructured China's government and economy.

Additional millions have since died in Viet Nam, Cambodia, Uganda, Rwanda, Sudan, Iraq, and Bosnia. However disturbing these last examples may be, the percentage of people deliberately killed has declined sharply in the past generation. Most countries are becoming more populous, many have are now more democratic, and more respond to international pressures to respect human rights, as outlined in the United Nations Charter. Each nation also depends on others for energy, raw materials, factory products, and markets, so wars and aggressive policies are increasingly impractical. The carnage and destruction of two world wars and the development of nuclear weapons demonstrate the folly of large-scale warfare and armed conflicts of the past half century, although numerous, have been relatively confined in area and scale.

Migration to New Areas

Extensive migration has facilitated large increases in total world population throughout history. When large numbers of people leave crowded areas, the ratio of remaining inhabitants to resources improves and the death rate often declines. Most of the millions of emigrants to sparsely populated lands, such as the United States and Canada a century ago and to Siberia more recently, remained in the new lands. Many were young adults who found largely untapped resources, such as minerals, timber, cheap farmland, fish and game, fresh air, and abundant water, and used them to support new families.

Ireland had a population of about 8.5 million when the potato blight struck in the 1840s and at least a million people died in the famine that followed. By 1854, 1.6 million Irish had emigrated to the United States. Today the population of Ireland is a scant 3.6 million, whereas 38 million Americans reported Irish ancestry in the 1990 census. Clearly emigration permitted the world total of Irish to increase substantially and this same principle applies to emigrants from other countries who found other lands with more abundant resources.

Around 1820, at the beginning of the modern population explosion, the present area of the United States supported about 12 million people. During the Great Migration from 1820 to 1930, about 38 million people, mostly from Europe, migrated to the United States and additional millions moved to other sparsely populated areas. Ultimately some of them returned home or migrated elsewhere, but the net gain, with additions from subsequent births, brought the U.S. population to 123 million in 1930 and the population has more than doubled since then.

International migration is no longer a promising solution to global overpopulation because the total population is now much greater, almost all of the habitable areas have been populated, and most countries have quotas that limit total immigration to a small fraction of those who would like to resettle.

Explaining Slow Growth

Large families were once the cultural ideal in most societies and were needed to maintain a stable population in the face of a universally high death rate. As noted above, the current surge in population growth in much of the world is due to the persistence of this tradition in a time of rapidly declining death rates. Families of 6 to 10 children are still common in some developing nations, especially among rural people, and in a few ethnic groups of the U.S. and Canada.

Theory of Demographic Transition

Today, average family size in the established industrial nations is from 1.5 to 2.0 children and very large families are rare. If the large family tradition had persisted until today, these same countries might now have two or three times their present populations. Because of their low death rates, they could be growing faster than Africa and south Asia, assuming that they were able to maintain reasonably high living standards with so many additional people.

The seemingly universal decline in population growth rates in industrial countries is addressed by the Theory of Demographic Transition. This theory explains that before becoming industrial, every country needed high birth rates to offset high death rates. As industrialization progressed and people were better educated, each population grew rapidly because of reductions in the death rate. Once large-scale industrialization was achieved, the birth rate also declined.

Why Fewer Births?

The decline in fertility in industrial nations coincides with the emergence of a well-educated middle class that sees the relationship between small families and a better quality of life for women and their children. Industrial countries accord women a higher status than formerly and increase their options by providing career ladders and other goals outside of the family circle. Parents are willing to substitute their own judgment for traditional expectations about family size and the role of women, and have access to family planning information and effective birth control techniques.

Most families in industrial nations no longer perceive children as an economic asset. They live in urban areas and depend on cash earned outside of the household to meet their material needs, including housing, food, clothing, transportation, schooling, recreation, and health care. In addition, school attendance is required for all children, their labor is restricted, they are increasingly expensive to support, and they are difficult to control because of outside influences.

Outlook for Developing Countries

Some population experts think that large families in populous developing nations with limited resources will "automatically" diminish in the near future without resorting to strict, government-imposed family planning programs like the one in China. Others think that more initiatives to increase the rights of women in these countries will help to reduce family size. Optimists cite the examples of Taiwan, Thailand, South Korea, and Singapore, where industrialization is now well established and birth rates have fallen rapidly without extreme coercion. Others disagree, believing that the situation in many of the less-developed countries of Asia, Latin America, and Africa is quite different from the Western experience and therefore the outcome may also differ. In this view, the Theory of Demographic Transition may not apply as well to these nations. Western medical technology and influence already have helped to reduce the death rate in developing countries worldwide, without an equivalent reduction in the birth rate in most cases. Many nations now have huge populations in relation to their productive land area and natural wealth, and are struggling to meet the basic needs of their citizens (Ch. 4, 7). Achieving full-scale industrialization, full employment, a large middle class, and higher living standards are formidable challenges for many of them. To compete in today's world, a country needs adequate capital investment, available natural resources, an educated labor force, economic options, and political stability.

Chapter 3

Escalating Consumption of Resources

Historians and others sometimes label various periods of history to reflect dominant themes, e.g., the Age of Faith, the Age of Exploration, the Age of Enlightenment, the Electronic Age, etc. An apt title for the past half century would be the Age of Consumption. Modern industrial societies produce and consume an ever-expanding array of goods and services, use energy and raw materials at an unprecedented rate, and increasingly rely on imports to meet their needs.

In most complex societies historically (and in some today), a small minority of people enjoyed power, privilege, and prosperity while the remainder struggled to survive as peasants, slaves, laborers, nomadic herders, soldiers, or small shopkeepers. Thus, average per-capita consumption of many natural resources was low by modern standards. During this century the situation changed dramatically in the highly industrial nations where the majority of citizens are now zealous consumers of a wide range of factory goods and commercial services.

Meanwhile "developing" nations strive to surmount the obstacles of rapid population growth, cultural restraints, technological lag, and sometimes meager natural resources that limit their material living standards. China, with a population ten times greater than Japan and over four times the U.S., now rivals Japan for the globe's second largest economy and could overtake the U.S. within a few decades.

Presently about one-fourth of the world's people, a "consumer class" of about 1.5 billion people, buys the majority of the planet's commercial goods and services. Although some of this class exists in every country, it varies in size from most of the residents of some prosperous industrial nations to a small fraction of people in countries where ancient traditions prevail, natural wealth is limited, subsistence agriculture is the mainstay, and many people lack land and jobs.

The Industrial Nations

A country's total industrial output is a rough measure of its consumption, since a large share of its goods and most of its services usually remain within the country. The monetary value of all of the goods and services a nation produces at home is called its gross domestic product (GDP). Its gross national product (GNP) is calculated by adding net income from foreign sources to GDP. In the U.S. GDP was 98 percent of GNP in 1993.

GDP is a popular measure of a nation's industrial activity and its rate of economic growth, while per-capita GDP provides a clue to its prosperity. It is an incomplete measure because it does not include nonmarket activities, such as goods and services produced by families for their own use. GDP considers the cost of government services but not their perceived value to the society. It does not debit unpaid costs, such as the loss or degradation of resources by industrial activities or related human suffering. Neither does it tell us how equitably these goods and services are distributed among the populace or regional differences in their actual cost to the consumer. But even with these limitations, GDP is useful for depicting trends in natural resource use and for comparing per-capita consumption in different nations.

Leading Consumer Nations

In 1993, about 20 nations (plus the small principalities of Europe), with just 12.3 percent of the world's people, had GDPs of over $16,000 per-capita. Annual population growth rates in these countries are between 0.0 and 1.4 percent, and almost all are well below the global average of 1.6 percent. These highly productive industrial nations include:

Australia	Finland	Japan	Sweden
Austria	France	Luxembourg	Switzerland
Belgium	Germany	Netherlands	United Kingdom
Canada	Iceland	Norway	United States
Denmark	Italy	Singapore	

About ten other industrial nations have rising levels of industrial production and per-capita GDPs from $6000 to $14,000, including Argentina, Greece, Ireland, Israel, New Zealand, Portugal, Korea, Slovenia, Spain, and Taiwan. Another ten, Brazil, Chile, Malaysia, Hungary, Poland, Estonia, Russia, the Czech Republic, Mexico, and Uruguay, range from $3000 to $6000, still well above the world median of about $1200. Several nations with unusually high oil export income per capita, such as Kuwait and the United Arab Emirates, have high GDPs despite limited production of factory goods. A few other small nations do well as tourist meccas.

Thus, using the GDP measure, about 50 nations lead the world in their per-capita production and consumption of factory goods, commercial services, and raw materials, such as minerals, fossil fuels, timber, farm products, and water, and about 20 of them use much more than the others. The remainder, including most of the world's most populous countries, have much lower GDPs per capita, the majority of them from $100 to $1000. Although the annual GDP of the poorer half of the world's nations is now increasing faster (4.6 percent) than in the richer half (3.2), their per-capita gain is less (2.7) due to rapid population growth (1). Annual growth rates in low-GDP nations usually are between 1.5 and 4.5 percent, so these have a high percentage of children in the total population.

The situation in many developing countries is not as desperate as these figures suggest because GDP statistics do not include the value of most family maintenance activities, such as subsistence farming, homemaking, gardening, fuel and food gathering, making clothing and furniture, non-commercial home renovation and repairs, and bartering. All of these tend to be important means of livelihood in developing countries, especially those with large rural populations. In contrast, most families in urban industrial countries produce relatively little in their households and are much more dependent on commercial goods and services, thus increasing their GDPs. Many commodities on the world market have been getting cheaper (in constant dollars) due to greater efficiency in production, reductions in tariffs, improved transportation, and fluctuations in the value of currencies. This encourages more consumption. In addition, the currencies of some developing nations are undervalued on the international market and will purchase more locally than the exchange rate suggests.

The United States, with 75 persons per square mile, and Bangladesh, with 2478, well illustrate the polar extremes of consumption between industrial nations and others with very limited industrial facilities. Per-capita gross national product, commercial energy consumption, and average income in the U.S. are about 100 times greater than in Bangladesh, while water use is tenfold, daily calory consumption is almost double, adult literacy is twice as high, and life expectancy is 25 years longer (2). Despite its severe natural resource limitations, the material needs of Bangladesh are increasing rapidly because its population is growing 2.3 percent annually compared to 1.0 percent in the United States.

Goods and Services Consumed

Another basis for comparing both domestic and international consumption patterns and trends is the purchase or use of specific goods and services. For example, one can compare the number of registered autos and television sets, total highway mileage, meat consumption, calorie intake, and educational attainment. Citizens in the United States and other industrial nations lead the world in their demand for living space, autos, home appliances, clothing, prepared foods, commercial recreation, health care, schooling, toys, and many other goods and services. Most people in developing nations enjoy far fewer goods and services. In 1992, the number of persons per registered passenger car was just 2 in most industrial nations, but it was 7 in Argentina and Bulgaria, 22 in Costa Rica, 50 in the Dominican Republic, 65 in Thailand, 177 in Syria, and even higher in many other countries. In 1991, the number of television receivers per 1000 people was 450 to 800 in most industrial countries, but 207 in Brazil, 175 in Turkey, 74 in Algeria, 59 in India, and 21 in Guatemala (3).

A good general measure of consumption is commercial energy use because so many other types of consumption vary with it. Energy consumption in the United States almost doubled between 1960 and 1992, Europe more than doubled, Asia increased sixfold, and total global use almost tripled with per-capita consumption rising 50 percent. Oil was the leading source of energy (37 percent), followed by coal (25), natural gas (23), hydropower (6), and nuclear power (6) (4). Alternative forms of energy, such as solar, wind, and tides, are difficult to assess, but are still relatively minor sources (Ch. 11).

The United States as Role Model

Foreigners often marvel at the high living standard of the American people as revealed in films, on television, and in statistics such as disposable family income, home ownership, and automobiles per family. This standard, in part illusory because it is beyond the reach of almost a third of Americans, is based on material possessions more than on overall quality of life. Today this standard is increasingly evident and often more equitably distributed in other industrial countries.

Webster defines standard of living as "the necessities, comforts, and luxuries enjoyed or aspired to by an individual or group." The United States is indeed at or near the top when the total consumption of goods and services is the primary measure of this living standard. Only a few countries rival this nation in per-capita (1) disposable income (2) private ownership of spacious homes, yards, cars, swimming pools, appliances, telephones, and other personal property, (3) availability and use of cheap energy, water, and construction materials, and (4) the size, number, and appointments of their school buildings, hospitals, shopping malls, libraries, parks, playgrounds, and other public facilities. The extent of this consumption is summarized as follows.

Energy Use

In 1993, per capita commercial energy consumption in the United States was about 5.4 times the global average. Renewable energy (hydropower, wind, solar, geothermal, and wood) was estimated at 9 percent of the U.S. total. Per-capita energy consumption is 1.6 to 2.3 times greater than most European nations and Japan, 6 times Mexico and Iran, 13 times more Brazil and China, 31 times India and Morocco, and over 100 times Kenya and Haiti.

A portion of these differences can be attributed to extensive industrialization and colder winters in the U.S. and to the greater use of noncommercial fuels (wood, peat, or cattle dung) in developing countries. But the much higher rates of consumption also reflect a reliance on private autos rather than mass transit, the large single family homes, processing virgin materials rather than recycling, maintaining constant indoor temperatures, a remarkable array of conveniences, extensive use of expendable products, and excessive waste from careless use. Visitors often observe that Americans seem to overheat buildings in the winter (rather than dress warmly) and overcool them in the summer (despite lighter clothing). Prior to World War II, wood and coal were the most widely used fuels in the United States. Oil and natural gas were also abundant and surpassed coal consumption in 1950 (oil) and 1958 (gas). Major new discoveries abroad in the 1940s and 1950s greatly increased the world supply of petroleum, reduced the price, and by 1947 the United States was a net importer of oil (4). Dependence on foreign sources increased to about half of U.S. oil needs by 1995. These imports, along with cars, industrial equipment, electronic appliances, clothing, shoes, and assorted gadgetry, created a large and persistent negative balance in trade between the U.S. and several other nations.

Automania

Much is said about the American "love affair" with their automobiles, a romance that is encouraged by an extensive network of paved roads, the popularity of suburban living, minors driving and owning cars, free parking at malls and "strip cities," and remarkably inexpensive fuel. In 1992 Americans had 190 million registered motor vehicles and 144 million were cars (6). Thus the entire population could be riding at the same moment in cars alone without even using the back seats. The U.S. leads all other nations in distances traveled in private automobiles and aircraft while, due to reduced demand, interstate railroads and buslines now provide only limited service between major population centers.

Cars account for 85 percent of passenger miles in the United States and one of six American workers is said to be employed in an auto-related industry. More land (an area the size of Georgia) is allocated to streets, highways, and parking than to housing (7). Gasoline and diesel fuel cost one-third to one-half as much as in most of Europe, India, Japan, and many other countries where high prices limit the ownership and use of private autos, while also ensuring a market for public transportation and reducing pollution and traffic congestion that would otherwise occur. In the U.S., 85 percent of commuters drive alone to work and each year spend billions of hours in slow traffic with frayed nerves, wasted leisure time, and idling engines that consume fuel and emit noxious gases (Ch. 5).

Demand for Goods and Services

Historically, consumption in the United States, Canada, Australia, and a few other countries has been encouraged by the abundance of natural resources in relation to population, relatively unfettered free enterprise systems, liberal federal resource development policies, easy credit policies, and the high aspirations of many immigrants. Even today, land, cars, appliances, clothing, foods, and many services in the U.S. are inexpensive in relation to family income and their cost in other industrial countries.

Being hard-pressed financially is of course a relative situation and one hears how difficult it is for an average family to make ends meet these days. Yet according to the U.S. Bureau of Economic Analysis, American's per-capita expenditures for consumption (in constant 1987 dollars, eliminating the effect of inflation) have increased from $5107 in 1945 to $8842 in 1970 and then to $13,716 in 1994 (8). Rising incomes, new types of products, massive advertising, expanding notions of need, more two-income families, fewer children, and perhaps liberal credit policies have made these increases possible.

Thus, even though the rate of consumption of some commodities has slackened in recent years, the general trend per-capita since 1950 has been clearly upward. Compared to a generation ago, Americans own 80 percent more cars and travel ten times as far by plane. The annual distance each person travels by auto is almost one-half more than the Germans, nearly double the British, over three times the Japanese, and 50 times the Thais (9).

No doubt the personal consumption of clothing, fast foods, prepared foods, electronic equipment, commercial entertainment, and many other goods and services is also much greater than in 1950. Many more homes now have two-car garages, television, air conditioning, freezers, dishwashers, clothes dryers, and power lawnmowers. Over 98 percent of homes have television, an average of 2.1 sets per home, most homes receive one of 65 million newspapers sold daily, and radio ownership averages 2.1 per person. Add to this a host of other products not generally available in 1970, such as computers, videotape recorders, microwave ovens, Jacuzzis, spas, dustbusters, and automatic garage door openers.

Domestic fresh water consumption by agriculture, industry, institutions, and households amounts to 1300 gallons per capita daily (10). Food is inexpensive, costing only 18 percent of family income in 1991 compared to 16 percent in France, 19 in Germany, 33 in Greece, 34 in Venezuela, and 53 in India (11) Over a million new housing units are begun each year, requiring extensive earth removal, vast quantities of wood products, concrete, asphalt, metals, ceramics, plastics, paper, paint, and other products. Hundreds of thousands of office buildings, industrial plants, institutional complexes, airports, roads, and utility corridors are also constructed, while others become dilapidated after one to three generations of use and must be renovated or razed. Obsolete buildings and facilities, along with junked cars and appliances, pose a major challenge for prudent waste disposal.

Americans often aspire to the lavish consumption standards of heirs to great fortunes, corporate executives, millionaire entertainers and athletes, and even lottery winners. Although a class of very wealthy people exists in virtually every country, however poor the great majority of citizens may be, it is unusually large in the U.S. Since 1970, the income gap between rich and poor Americans has been increasing. In 1993, 40 percent of households earned less than $25,000 annually, the same as in 1970 (in constant dollars), and apparently more people are now homeless. But the percentage of households earning over $75,000 almost doubled during this period (12).

Consumption in Developing Countries

Roughly 80 percent of the world's people live in developing countries. In rapidly developing nations, most of the people are keenly aware of Euro-American consumption patterns through the mass media and visiting tourists, and wish to improve their own material living standards. Many of these countries offer an expanding market for everything from industrial plants, mine equipment, hydroelectric dams, utilities, and highways to schools, hospitals, houses, cars, appliances, and furniture. People in nations with traditional subsistence economies would also like to have more jobs, goods, and services but often lack the technology, expertise, capital, natural resources, and/or political stability needed to achieve these goals for their rapidly growing populations.

If massive poverty is overcome, and that is the objective of various foreign aid programs, we can expect rising consumption in these countries as well. The world's population doubled between 1955 and 1995 and could double again in about 50-60 years, so the potential for future consumption is awesome, as the example of motor vehicles illustrates. If all nations owned as many motor vehicles per capita as Americans and western Europeans, the world would have nearly four billion vehicles. The planet now has over one-half billion cars, plus additional trucks and buses, and there are brisk sales of new cars each year.

It is realistic to expect a billion vehicles on the road within a generation (13). Traffic jams, parking, and exhaust pollution already are serious problems in many parts of the world, so one can imagine the implications of doubling or quadrupling the vehicle population without compensating pollution control and land-use planning to minimize converting valuable agricultural lands to roads, strip cities, and parking lots. Rapid population growth in most developing countries accounts for just a portion of their rising levels of consumption. Many countries are consuming more per capita due to extensive new construction, increased domestic output, and the payrolls these activities create. Since World War II, most Asian, Mideastern, Latin American, and many African nations have erected and equipped new factories, built new roads, developed additional raw materials, expanded their cities, and increased exports of their factory products.

The percentage of economic growth and increased per-capita consumption in several successful developing nations is much greater than in the industrial West. Taiwan, South Korea, Singapore, Malaysia, Thailand, and China have developed a lively export trade in textiles, appliances, and a host of other products. Saudi Arabia, Kuwait, and the Arab Emirates use some of their vast oil revenues to improve social conditions.

Total world demand for minerals and mineral fuels has climbed 50 to 75 percent each decade since 1950, but population growth accounted for less than half of this increase. Per-capita consumption approximately doubled for energy, copper, steel, timber, and meat, quadrupled for cement and car ownership, quintupled for plastics, septupled for aluminum, and increased even more for air travel, air conditioning, television ownership, and numerous other goods and services (14). To grasp the overall impact on world resources, remember to double these figures because the planet's population has doubled along with per-capita consumption in this span of time.

The upward surge in consumption, evident in virtually all countries, has benefited some people much more than others. Living standards for the poorest third of the world's people, most of whom live in developing countries, have not increased very much during this time. Millions of people lack shelter, verge on starvation, lack education, jobs, and medical care, and would need to consume much more to reach the living standards of most welfare recipients and prison inmates in the industrial nations (Ch. 7).

Industry's Role in Consumption

The quest for profits is the primary goal of a business or industry, even though secondary objectives such as product quality, employee welfare, and environmental protection are also important to some firms (Ch. 6). Profits depend on sales volume and productive efficiency by cutting costs, reducing the work force, using the cheapest acceptable materials and processes, and minimizing responsibility for environmental impacts.

Cost cutting often results in decisions that provide short-term benefits for a company, its current managers, and its investors but may also be inconsistent with long-term needs of the firm and the best interests of its host community or nation. This is evident in the earlier "reap, ruin, and run" practices of the mining, petroleum, timber, livestock, and fishing industries that harvested the cream of the nation's resources, moved on, and now sometimes anguish over future supplies (Ch. 4-5).

Strategies to increase sales include introducing new models or styles annually or seasonally (planned obsolescence), creating new products and touting them as improvements over those already available, and the use of expensive, eye-catching, or merely convenient packaging that soon becomes solid waste. Thousands of new products appear on supermarket shelves each year, many similar to earlier versions but more likely to increase total sales. Some firms offer products that look good and cost little but wear out rapidly or are hazardous to use. Poor quality often costs consumers more in the long run, generates excessive pollution and waste, and may lead to government-imposed measures to protect the public interest (Ch. 4-5).

Some new products are hard to justify from an environmental perspective but are nevertheless popular and profitable for their suppliers. A good example is the disposable beverage container introduced in the late 1960s, a time when cities were becoming very concerned about the growing volume of solid waste and the shrinking acreage available for its disposal. Prior to this, most beverages were sold in bottles that were routinely returned and reused. Although some states now require deposits on certain cans and bottles, container companies and retailers strenuously oppose legislation to extend this practice to other states or to additional containers. Trillions of containers still end up as litter on streets, lawns, and highways, and as municipal solid waste after just one use.

America's commercial interests use a remarkable variety of techniques to sell their products and these costs are passed on to the consumer. Americans spend at least 55 billion dollars ($210 per person) annually advertising goods and services, the majority of it on newspaper, television, and direct mail ads. (15). Both massive advertising and the ever-increasing variety of new products and services stimulate consumption by increasing the threshold of need.

Each year fewer adults are willing to forego air-conditioning in their home, office, and auto, or to get by without a second family car, television set, or novel new appliance. Today's children consume more clothes, toys, electronic equipment, and junk food containers than their parents at the same age, far more than their grandparents, and carry these higher expectations into adult life.

Some business and industrial firms spend additional millions of dollars trying to convince the public that we have ample natural resources, that environmental degradation is under control, that controversial products are safe, that business taxes should be reduced, and that political candidates who support free-market policies should be elected (Ch. 10). Do business leaders favor population growth because of its potential for low-cost labor and increased product sales. Their critics, including numerous environmental groups, consumer advocates, and other interest groups, challenge these assumptions (Ch. 9).

In the short run, optimal consumption and waste are good for business, often increasing both production and profits. In the long run, they deplete our nonrenewable resources, encourage reckless natural resource development, jeopardize vital ecosystems, and endanger human health. As we move closer to a world community of shared resources and opportunities, the stark contrast between have and have not nations and the rich and the poor within nations becomes more apparent and harder to justify morally. Yet many of the people who prosper under the present system want to preserve it as long as possible.

Standard of Living Reconsidered

Media reports indicate growing national and international concern about natural resource depletion, adverse impacts on the environment, and the welfare of future generations. We read that industrial societies have become very materialistic and need to consider alternative measures of well-being. Critics charge that much of what we buy is relatively nonessential. Some families have more cars than drivers and others waste a third of the food they buy. Perhaps many consumers would be healthier and happier without so many goods to maintain or store (displacing cars in their garages), rich foods to clog their arteries and make them overweight, and conveniences to deprive them of needed exercise.

Redirecting Values

Any serious effort to protect the environment and to conserve resources for future generations must grapple with the problem of mounting consumption, which rivals or exceeds population growth in depleting natural resources and degrading the environment. Many approaches are possible and some are widely used, including family planning, sustainable use of renewable resources, efficient use of nonrenewable resources, extending the life of products, recycling when feasible, and learning to live with less (Ch. 11).

Many people do seem to think that personal happiness, social status, and career success are best validated by numerous possessions and a lavish lifestyle, and that friends, relatives, and co-workers are best impressed by displays of affluence. There is some evidence to support these contentions. Even though the average family's purchasing power has increased considerably since 1960, more families are using multiple credit cards to the limit and the number of bankruptcies has tripled since 1981. Some social critics think that our emphasis on materialism results in an erosion of traditional family and spiritual values, alienation from community life, and increases in antisocial behavior such as drug abuse. They urge us to reexamine our values and to improve our quality of life by focusing on satisfying alternatives to consumption. Many enjoyable activities are relatively nonconsumptive and healthy as well.

Quality of Life Redefined

With the sharp decline of the dollar compared to other currencies since the 1970s, several other industrial nations have forged ahead of the United States in per-capita income. Often their consumption of material goods remains lower because they have less land available for development, pay higher taxes, save more money, value different things, and pay more for their goods and many services than in the U.S.

If standard of living is defined as overall quality of life, quite a few countries exceed the United States in key criteria, such as average life expectancy, physical fitness, absence of poverty, freedom from serious crimes (such as murder, rape, robbery, assault, and drug abuse), feeling secure in strange neighborhoods, access to affordable health services, family viability, school achievement levels, and the absence of neglected buildings, litter, and graffiti. Despite the high levels of product consumption in the United States, comparative studies of poverty suggest that we have two or times the percentage of poor people, including children and the elderly, as other industrial nations, such as Canada, Germany, and France.

The millions of dollars Americans award to business executives and entertainers, to athletes and film stars who endorse products, and to organized criminals for illegal services attest to the effectiveness of advertising techniques and to the importance of short-term gratification in our lives. America's families, educators, and public officials have a responsibility to define and pursue long-term goals that will guide emerging generations and help to ensure an acceptable quality of life in the decades ahead.

Part Two
Environmental Impacts of the Trends

Chapter 4

Depletion of Natural Resources

In the 1960s, economist Kenneth Boulding compared our planet to a giant spaceship on a long voyage. Like any other space vehicle, "Spaceship Earth" has finite resources that must support its passengers until the voyage ends and it is a very long trip through countless generations of time (1). When Earth's burgeoning population runs out of mines and sewers, Boulding explains, everything will have to be recycled. Spaceship Earth became a popular analogy to help people understand the ultimate limitations on resources if future needs are considered. Today we know the situation is further complicated by other factors, such as ocean pollution and overfishing, global warming, and plant and animal extinction (Ch. 5).

More than half of all human consumption of many important resources, such as iron, copper, oil, gas, pulpwood, fertilizers, cement, and artesian well water, has occurred within the past three or four generations. Because the population in most countries is increasing, we can expect even greater demands until the nations of the world stabilize their growth and reduce their consumption of virgin (newly extracted) nonrenewable and hard-to-renew resources. Many scientists and conservationists now urge all countries to develop sustainable economies that conserve, renew, and recycle resources and protect environmental quality (Ch. 8). They believe this formidable challenge must eventually be met if humanity is to survive and prosper.

Vast quantities of natural resources and synthetic materials (most derived from these resources) are needed to maintain the high and rising material living standard of the industrial nations, as depicted in Chapter 3.

According to Young and Sachs, leading industrial countries with about 20 percent of the world's people consume 80 percent of its iron and steel, 86 percent of its aluminum, 81 percent of its paper, and 76 percent of its timber (2). Of course all of these nations export some of the products they manufacture from these materials but they also import factory goods.

Per-capita rates of consumption of raw materials and factory goods in the more successful developing nations are now rising even faster than in Europe and America, and most other countries want to increase theirs as well (Ch. 3). Even where economic growth is sluggish, rising demand usually occurs due to rapid population growth.

Renewable and Nonrenewable Resources

Resources such as industrial minerals, metals, and fossil fuels, topsoil, wilderness areas, and extinct species are often called nonrenewable because once they are used up, substitutes must be found. Renewable resources, such as solar and wind energy, forest products, food crops, fresh water, animals, and "tired" croplands, usually can be renewed, reclaimed, or reproduced when economically feasible.

The renewable versus nonrenewable distinction is sometimes difficult to make. For example, metals are renewable if they are recycled (most gold and much silver are) whereas plants and animals are nonrenewable if they become extinct or their numbers are too few to reproduce viably. Fresh water from rain is regularly replenished but when deep aquifers containing water accumulated for thousands of years are drained, new sources must be sought. Soils can be enriched by adding nutrients if they are merely depleted from overuse, but soils lost to wind and water erosion or other land uses are very expensive to recreate.

Modern land-use policies and technology permit the extraction and use of minerals, fossil fuels, timber, water, and other resources, often at bargain prices that do not reflect all of the related environmental costs. These "external" costs include air and water pollution, soil erosion, ecosystem disruption, social impacts, and the disposal of discarded products (Ch. 5). Publicly-owned natural resources (e.g., minerals, timber, and croplands) are often sold at prices so low that governments collect little or no revenue and may even lose money on the sales transaction.

We can best evaluate the current U.S. and global situation by reviewing the nature and importance of various natural resources and the demand for them in relation to estimated supplies.

Metals and Industrial Minerals

Minerals, broadly defined, include any of the materials removed from the earth for industrial or home use, including metals and their ores, construction materials, other industrial minerals, and various fossil fuels and gases. Iron, aluminum, and copper are the most widely used metals and several more, including zinc, lead, nickel, manganese, magnesium, chromium, tungsten, titanium, and tin, are used to make alloys of these metals that are stronger, harder, more corrosion resistant, and/or can withstand great heat.

Zinc, lead, and mercury have many additional uses, as in batteries, electrical equipment, pigments, and various chemical compounds. Silver, platinum, and gold are used in coins, jewelry, and dentistry, and have many other industrial applications. Yttrium, thorium, and other rare earth metals are used in electronic equipment and their oxides have additional uses. Uranium is the source of fuel for atomic reactors.

Industrial minerals such as granite, cement, gypsum, limestone, various insulation materials, and gravel are indispensable in the construction industry. Other minerals are used for ceramics, abrasives, computer chips, fertilizers, pesticides, paints, jewelry, medicines, diet supplements, and as basic chemicals for laboratory work. At least 100 minerals have become vital to modern life and over 3000 have industrial uses.

New minerals products are developed each year and, in the modern era, the increase in demand easily exceeds the rate of population growth. From 1750 to 1900, the world's demand for minerals multiplied about tenfold while population only doubled. Since then, the rate of population growth has accelerated and the demand for minerals has increased thirteen times (3). Mines and smelters now consume about 10 percent of all commercial energy. After 1975, the rate of increase in the demand for minerals slackened, at least in industrial nations, due to increased energy costs (that also encouraged more efficient use), the substitution of plastics for metals, increased recycling, economic recessions, and other factors.

Worldwide, the total value of minerals and mineral fuels consumed annually is close to two trillion dollars. Massive amounts are needed and annual production now includes about 725 million metric tons of steel, 19 million of aluminum, 19 million of copper, zinc, and lead, and 3 million of chromium. Add to this 11 billion tons of stone, 9 billion of sand and gravel, 1.3 billion of hydraulic cement, a half billion of clay, a fifth billion of salt, and large amounts of phosphate, potash, lime, and sulfur (4).

A scarcity of mineral resources worldwide is not an immediate threat to future generations. Many minerals, such as aluminum, iron, calcium, magnesium, and silicon, are abundant in the earth's crust. Improved technology, including powerful explosives, huge trucks and bulldozers, and sophisticated mining machinery, now enable mining companies to process ores once considered too lean for profitable development (e.g., copper ore that is 99 to 99.5 percent waste). Of greater concern are the potential environmental impacts of ever-expanding production of virgin resources from increasingly marginal ores (Ch. 5).

Geologists in search of minerals have combed the globe and many if not most of the known, rich, easily accessed deposits of key minerals and mineral fuels are now being developed or are already depleted. Most newer metallic ore discoveries are deeper and less concentrated, so enormous quantities of earth must be removed to extract them, especially the ores containing precious and rare metals, such as uranium, platinum, and gold. At risk are ecosystems, air and water quality, wildlife habitat, underground water flows, and the health and quality of life of area residents and livestock (Ch. 5-7). Some minerals, including aluminum and rare earth metals, also require tremendous quantities of energy to mine and refine.

Minerals policies in many countries encourage tapping new resources more than conserving and recycling existing supplies. Some nations subsidize mineral exploration and development, make tax concessions to these industries, and are less stringent about environmental protection than with other industries. In the United States, under the 1872 mining law (intended to encourage development at that time), a person or firm (including foreign companies) can secure title (ownership) on most public lands to both a new ore discovery and the land above it for a nominal fee ($5 per acre or less) and is not required to pay royalties to the government for the minerals extracted. Worldwatch Institute reports that miners extracted $4 billion in minerals from former public lands in 1988 alone (5).

Under the 1872 law, an area about the size of Connecticut has already been purchased at bargain prices and, according to the Common Cause public interest group, 15.5 billion dollars worth of public lands with minerals under pending applications could also be sold to mining firms for less than one million dollars. Congress is reluctant to change the law because of strenuous opposition from mining interests and their political allies, and the desire to minimize dependence on imports of vital minerals.

Fossil Fuels and Gases

Fossil fuels, namely various types of coal, oil, natural gas, and peat, are the lifeblood of modern societies. We use them for mining and manufacturing operations, daily food preparation, producing most electrical energy, lighting, heating buildings, propelling vehicles, and as raw materials for thousands of products, such as lubricants, asphalt, tar, synthetic fabrics, plastics, cosmetics and creams, pesticides, and waterproof finishes. Coal, oil, and natural gas have largely replaced wood and peat fuels in industrial countries because they are cheap, easily transported and stored, burn more efficiently, are less messy, and produce higher temperatures.

World petroleum consumption increased dramatically each decade following the development of gasoline and diesel engines, oil heating, and paved roads. The rate of increase in consumption during the 1960s and 1970s was phenomenal until it leveled off in the 1990s at about 24 billion barrels (each holds 42 gallons) per year. No doubt the recent development of more fuel-efficient automobiles and appliances and improved building insulation helped to stablize consumption. The United States consumed over six billion barrels of oil in 1994, about 25 percent of the world total, and imported about half of this (6).

While per-capita consumption of commercial energy in industrial nations has tapered off since the 1970s, it almost doubled in developing nations, reflecting population growth, more industrialization, rising living standards, and inefficient use of energy. Most countries now import the bulk of their oil, mostly from a dozen nations with rich deposits that exceed their domestic needs.

Scientists believe that all of the fossil fuels are primarily organic in origin, created when ancient forests, swamps, and animal remains were buried and subjected to millions of years of intense heat and pressure. Thus they are limited in quantity, although extensive prospecting and improved technology have increased the available supply each generation. Many petroleum geologists think that world production will peak in another decade or two and then begin to decline as major fields are gradually depleted. Crude oil, currently in greatest demand, may be the first to be depleted because of limited supplies.

British Petroleum Corporation estimates that world oil reserves will last another 40 years at present rates of consumption, natural gas 50 or more years, and coal over 200 years. Some other estimates are more optimistic. Major variables are the volume of petroleum in untapped deposits, the extent of undiscovered sources, and the price of oil. Without imports, U.S. "proved" oil reserves would only supply its needs for only four years, but higher prices would encourage more drilling in marginal areas now considered unprofitable to develop. Known U.S. reserves of coal are far more abundant. High oil prices would increase their use and perhaps increase pollution as well.

The present "glut" of oil on the world market is mainly due to simultaneous development of most of the world's richest known deposits along with competition among a dozen developing countries to export their crude oil at low prices in exchange for the "hard" currencies of industrial countries. This enables them to import factory goods and food, and to finance their own industrial development (Ch. 1, 6).

No doubt more of each fossil fuel will be discovered before these known reserves run out but no one knows how extensive they are or how expensive they will be to extract and market. Some geologists believe that virtually all rich accessible deposits of petroleum have been discovered. But additional deposits probably exist at deeper levels in the earth or under the continental shelves.

When the richer oil reserves are depleted and the price goes up, petroleum companies can extract oil from abundant oil shale and tar sands, producing a great deal of waste material in the process. As with metallic ores and industrial minerals, the critical questions are: How much earth (and how many drainages and ecosystems) are we willing to disturb to get more fossil fuels? And how much are we willing to pay at the pump? In time the world will need to depend much more on alternative fuels, such as wind, solar, tidal, and possibly hydrogen or nuclear fusion sources.

The world consumes about 3.5 billion tons of coal annually, with the coal-rich U.S. producing 23 percent of the total, 90 percent for domestic use. Coal is used to produce electricity and steam power, for heating, and as a source of organic chemicals widely used in manufacturing paints, plastics, drugs, explosives, preservatives, roofing, and other products. Anthracite, the hardest, purest, least polluting coal is in short supply and hence expensive. Bituminous coal (softer, black) is much more abundant, as is lignite coal (brown, crumbly), which contains the most impurities. Peat is also composed of organic matter and is even less refined by nature than lignite. It is used as a garden mulch, for livestock litter, and as a cheap fuel in a few countries.

Natural gas is both a by-product in oil fields and found in separate deposits. It is a mixture of flammable gases, mostly methane, but also contains varying quantities of butane, propane, and other hydrocarbon compounds. With proven reserves eight times annual consumption in the

U. S., and more being developed, natural gas is still fairly abundant and increasingly valued because it pollutes less than other fossil fuels (7).

Together these fossil fuels supply most of the energy consumed in the United States. The major commercial energy sources, with the percentage each supplies, are oil (40), natural gas (23), coal (22), nuclear power (7), hydropower (3), and geothermal, solar, wind, and other sources (4) (8). Wood is still used for heating some rural areas. The current U. S. consumption of one-fourth of the world's commercial energy is down from one-third in 1970, due to better energy efficiency and increased consumption in developing nations (9).

Forests and Forest Products

Estimates vary (depending mainly on how forests are defined), but roughly 35-40 percent of the earth's land surface is forested (10). About two-thirds of this is in trees, including continuous forests, scattered groves, and tree plantations. Another third is in brush and other woody vegetation. The remainder of the earth is covered by great belts of grasslands, large expanses of desert plants, ice fields, and "badlands" of natural or human origin, where little vegetation is evident, often due to extreme aridity or to continuous erosion.

Vast slow-growing coniferous forests intermingled with deciduous groves are found at higher latitudes (e.g., northern Canada, Scandinavia, Siberia, and southern Chile) and at high elevations. Deciduous (hardwood) forest belts are common in the temperate zones of Europe, Asia, and the United States, while conifers (softwoods, especially pines, firs, and cedars) tend to dominate humid subtropical areas of the world and the western forests of the U.S. Lush broadleaf evergreen (hardwood) forests occupy the Amazon and Congo basins and the peninsulas and islands of southeast Asia.

Forested areas are vital, even to urbanites. They help to stabilize climate, purify water, and reduce flooding and erosion from heavy rains and snow melts. As the Forest Service's "land of many uses," motto suggests, forests provide lumber and other building materials, fuel, municipal water sources, habitat for game and fish, numerous pharmaceuticals, fruits, nuts, berries, mushrooms, and other edible plants. Wood and agricultural wastes (biomass) supply about a third of the energy used in developing nations, a major reason for deforestation in drier areas with sparse forests.

Worldwide, forests are the home of hundreds of indigenous peoples, the primary source of their livelihood, and often have spiritual significance for them. They are also valued for parks and scenic vistas, wildlife viewing, nature study, outdoor recreation, solitude, and relief from the heat.

Although some sources dispute this, in many countries forest acreage and the volume of available timber is much less today than two or three centuries ago, due to massive conversion to croplands, extensive cut-and-run timber harvesting, and the construction of water reservoirs, town sites, resorts, industrial facilities, roads, and utilities corridors. Conservationists allege that as much as 95 percent of the original or "old-growth" forests has been cut in the lower 48 states of the United States and virtually all is gone in western Europe. About 60 percent of Europe's original forest lands have been converted to other uses.

In the western states of the U.S., an estimated 90 percent of the old growth has been cut, most of what remains is publicly owned, and about a fourth of this is set aside in national parks and wilderness areas. Environmentalists would like to see the rest of these older forests protected as parks, designated wildernesses, game refuges, and research forests, so that genetic diversity, wildlife habitat, scenic beauty, and research opportunities will be preserved for future generations.

A segment of the timber industry disagrees, arguing that we need the additional timber and related jobs, that people are more important than wildlife, and that mature trees will eventually die and be wasted if not harvested. Since 1988, reductions in the timber available from federal lands due to environmental concerns have clearly hurt many small timber operations in the West that had become very dependent on federal timber sales. Often left unsaid is that in the past two or three decades many more timber industry jobs were lost due to the increased use of machinery in mills and logging operations than to reduced production. Between 1970 and 1988 lumber production increased about one-third and log exports doubled while timber employment declined (11).

Critics of the pre-1988 harvests on public lands allege that some timber companies were cutting federal and private timber at unsustainable levels and reductions were needed. Harvests included large tracts of old-growth forests with complex ecosystems. They also note that extensive private sector sales of U.S. logs overseas have deprived domestic mills and factories of high quality timber for producing wood products and for building construction, both of which provide many more jobs than logging operations. Logs are still exported abroad, but those from federal sources are now excluded.

The nationwide housing boom and the diminished supply of federal timber have steadily increased the price of lumber, encouraging greater use of waferboard, particle board, steel studs, and other substitutes. Lumber and paper imports from Canada are increasing, while log imports from other countries have generated concerns about introducing insect pests and diseases.

Other promising solutions to reduced supplies of timber and wood pulp for paper are planned or in practice. Federal and private reforestation programs for cut or burned tracts are underway. Farmers are planting fast-growing trees for early wood pulp production. Tree farms (or plantations) are becoming more common, with plots planted in different years to ensure a steady supply of timber in the future. These "even-aged" forests do not provide the ecological diversity of natural forests, but they do yield timber that is easy to harvest. Currently forested areas in some countries, including the United States, are increasing in extent and density due to conversion of marginal farm lands to forests, reforestation of previously barren tracts, the creation of tree farms or plantations for future harvests, and planting shelter belts on grasslands. According the U.N. Food and Agricultural Organization (FAO), tree plantations are increasingly common and offset about 12 percent of tropical deforestation and 20 percent of deforestation in all developing nations (12).

Wetlands

Wetlands are areas where the soil is saturated with water at least part of the year, as in swamps, marshes, tidal flats, and fringes of ponds. Each wetland supports diverse plant and animal life and provides habitat and breeding areas for migratory fowl. Wetlands also impede floodwaters, trap sediments, stabilize shorelines, support fish and shellfish, and reduce erosion. Some wetlands, such as the swamps of Alaska, Minnesota, Georgia, Florida, and Louisiana, are large enough to support complex ecosystems, but most are small and often widely scattered, as in North Dakota and Saskatchewan. Together wetlands occupy about six percent of the earth's land surface and are crucial to its ecology, even though rapidly declining in extent.

Many wetlands are being converted to agricultural uses or are lost by draining, filling, building channels, peat removal, water pollution, and other hazards. A well-known example is the Everglades Swamp in Florida, where human activities such as agriculture, subdivision construction, pollution, outdoor recreation, and even tourism are jeopardizing the viability of a large and very unique wetland ecosystem. Even though as much as half of the world's and America's wetlands may already be lost, some members of Congress recently advocated redefining the concept to make more wetlands available for development (13).

Plants and Animals

Humans around the world depend on thousands of species of plants, animals, and microorganisms, for their daily needs and many of these are essential to health and well-being. Animals, including birds and fish, provide meat, dairy products, eggs, clothing, fertilizers, and other commodities. They also haul burdens, carry messages, protect property, kill pests, contribute to medical research, and consume offensive wastes. They provide companionship for people of all ages and are the basis for many outdoor activities such as hunting, fishing, falconry, horse racing, and bird and animal watching.

Plants provide wood, fiber, and other useful materials (see Forest Resources, above), ornamental shrubs, a great variety of food, animal fodder, medicines, scenic beauty, flood control, and soil stabilization. They also replace the oxygen we breathe and burn, store carbon, and help to reduce the rate of global warming. Microorganisms raise our bread dough, culture our yogurt and beverages, fight diseases and pests, and decay our wastes.

Many wildlife populations are declining in numbers and more are becoming extinct. Human population growth and advances in technology (e.g., road and building construction, industrial operations, modern vehicles, air and water pollution, rifles, and poisons) have greatly intensified the rate of species extinctions. Authorities differ in their estimates of the extent of this trend, but tend to agree that it is occurring on a scale unprecedented since the demise of the dinosaurs some 65 million years ago (long before humans arrived). This trend will have serious repercussions for human welfare if it continues unabated (Ch. 5).

Domestic animal populations have increased along with human populations, in prosperous countries where meat and dairy products are important diet items. But horses and cattle are becoming a luxury in populous developing nations that cannot spare productive cropland for pasture. In the United States and some other industrial nations, concern about cholesterol has curbed the upward trend in red meat and egg sales. Meanwhile the demand for poultry, which is less costly, has increased and fish is becoming more expensive than red meat because the demand is growing while the supply is limited. Pet populations are increasing faster than human population in some countries and consume a greater share of food resources.

Croplands and Soil Quality

Soils, especially nutrient-laden topsoils, are an extremely vital resource because the plants they produce are the major food source of humans and their livestock, and also yield the oxygen we breathe. Rich soils take decades or centuries to develop by natural processes, but can be depleted in a few seasons by severe wind and water erosion, compaction from heavy vehicles, excessive fertilization, repeated plantings of the same crop, and accumulated salt or other chemicals. They may also be buried by mining or construction operations. Some fragile soils, such as those of short-grass steppes and tropical rainforests, are especially susceptible to erosion and depletion. Fortunately, soil can also be enriched or restored when the technology and skills are available and the owner can afford to do it.

The total area of the world's croplands has increased about five times since 1700, replacing forests, prairies, steppes, wetlands, irrigable portions of drier lands, and even lakes and ocean shallows. Large sections of the original forests and wetlands of the U.S., Europe, coastal Latin America, China, and a growing share of the world's tropical rain forests are now croplands, pastures, and orchards interspersed by communities and transportation corridors. Croplands now occupy about 40 percent of the planet's surface and, in most countries, further extension of croplands is limited by the terrain, the need to conserve remaining forests, the water available to irrigate additional arid lands, and the length of the growing season. The Green Revolution of the past three decades is now worldwide. Crop yields are boosted by planting new strains of grains and vegetables, and by applying large quantities of fertilizer, pesticides, and water. This new technology has saved millions of lives and permitted some additional population growth, but agronomists fear that these practices may not be sustainable (Ch. 7). Global warming may eventually increase the growing season in northern and mountainous climates, such as Canada, Siberia, and the intermountain plateaus of western United States, so that new areas can be farmed. Meanwhile, oceans rising from melting ice deposits would flood some currently very productive coastal plains while portions of other croplands could become more arid.

Fresh Water Sources

Modern civilization requires enormous amounts of fresh water for agricultural, industrial, recreational, and household use. World demand for fresh water has tripled since the 1950s and the supply in some regions, most notably Africa, is now inadequate. The demand for water increases because of population growth, more extensive irrigation, and expanded industrial use, while the supply in many areas gradually diminishes (Ch. 5). Fresh water is obtained from natural lakes, streams, and reservoirs, from shallow wells that access groundwater collected in surface gravels and soils, from deep artesian wells that tap into layers of saturated rock and gravel (aquifers), and from costly desalination plants that process ocean water. Due to increased consumption of fresh water, deforestation, wetland drainage, and other factors, the level of surface water (the water table) is dropping below well depths in many localities. Some of the deep aquifers that took tens of thousands of years to fill are losing volume and may eventually be unproductive. The Ogallala Aquifer, the largest in the U.S., supplies eight states from South Dakota to Texas and is now being drained faster than nature can replace it.

Water shortages are most critical in heavily populated drier areas, such as the margins of the Sahara Desert and in southern California and Arizona. Such areas rely heavily on deep artesian wells or divert stream flows to supply cities and to irrigate orchards, fields, lawns, and gardens. The demand for Colorado River water in Arizona and California is now so great that the river no longer reaches the ocean.

Today 17 percent of the world's cropland or about 600 million acres, are irrigated and produce about one-third of the total food supply (14). As available agricultural land and water sources dwindle, it is difficult to increase irrigated cropland. Ocean water can be desalinated by distilling, electrodialysis, filtering, and other means but these processes require considerable energy. Desalination is not yet an option in many parts of the world because the water produced is too expensive for general use. This process is used in very arid areas such as Saudi Arabia where fresh water sources are no longer adequate and energy is cheap and abundant.

U.S. and Canadian per-capita demand for water is almost double that of other industrial nations, assumedly because of extensive irrigation and more numerous industrial and household uses, such as family swimming pools, hosing streets and driveways, daily showers, automatic washers, continuously running water while washing cars, and neglecting water leaks.

Americans consume an average of 1340 gallons of fresh water daily per capita and about a third of it is used to irrigate crops. Around 80 percent of it is from surface sources and the remainder is groundwater. Water consumption varies considerably from one place to another, from as little as 200 to 300 gallons per capita daily in some New England states to over

10,000 gallons daily in some western states that rely heavily on irrigation for crops (15).

Fisheries

Where available, fish have always been a popular food for humans, other mammals, raptors, and sea creatures. Most of the world's people live near bodies of water, and fish and seafood are a vital part of their diet, especially in Asia and Africa. Many valued species of wildlife would not survive without the earth's fisheries. The world's commercial fish catch in 1992 was 98 million metric tons, around five times greater than in the 1950s. With more people, increased urbanization, more polluted lakes and streams, and fewer local fishing options, the world is now more dependent on the commercial ocean harvest.

The ocean catch, which many believe now exceeds sustainable levels, is beginning to decline despite the inclusion of additional species. Food fish are not uniformly distributed throughout the oceans and seas (as sometimes believed) but are concentrated in the shallower waters over continental shelves, above seamounts, and around islands.

Commercial fishing today is dominated by thousands of giant factory trawlers, many a hundred meters long and electronically equipped to locate schools of fish. These ships are capable of catching a half million pounds a day, are gradually displacing smaller operations, and pose the threat of unsustainable fishing. According to a new study of the United Nations Food and Agricultural Organization, 70 percent of commercial species worldwide are either depleted, fully exploited, or recovering due to reduced fishing. A 30-million ton shortage is foreseen by the year 2000 unless this trend is reversed (16).

In the United States, both fish and seafood consumption increased steadily, from 11.4 billion pounds in 1980 to 20.3 million in 1993 (78 percent), much faster than population growth (13 percent) (17). One reason for this upsurge is the medical evidence that fish and many seafoods are much lower in saturated fats than red meat. Limited availability of popular food fishes, such as coho salmon, Atlantic cod, haddock, flounder, and lobster has driven up the price and stimulated sales of other species, such as pollock, mackerel, pilchard, and squid, and has also increased imports. Until 1985, Americans used nearly as much fish for industrial products (e.g., meal, oil, animal food, and bait) as for food but now they eat almost 80 percent of the total catch (18). As the demand and price rise, farmers are motivated to raise trout, catfish, and other species in private ponds as a cash crop.

Unique Natural and Cultural Sites

Most countries seek to protect and to preserve some of their heritage, including sites of outstanding geological, biological, archaeological, and cultural significance. These may be ancient ruins, unique buildings, religious shrines, monuments, older sections of cities, unusual geological formations, or areas of great scenic beauty. The United States has thousands of protected areas, such as state and national parks, frontier towns, graveyards, museum and library collections, game refuges, wild and scenic rivers, stately waterfalls, ocean beaches, and designated wildernesses.

Historically, as the population increased and became more mobile, more of these special places were set aside to accommodate growing numbers of residents and tourists. Further expansion is opposed by people who prefer private sector control of such resources or tax revenues from their earnings. National park and monument visitation in the U.S. increased from 66 million in 1985 to 87 million in 1994, up 31 percent in seven years (19). Visitor density is now a problem for some of these parks and other public areas. Yellowstone, Grand Canyon, and Yosemite National Parks are so popular that the influx of people and vehicles threatens both their ecology and the quality of the visitor's experience. Entrance is sometimes restricted and a policy of requiring reservations is being considered for peak seasons in some parks.

Outdoor Recreation

A growing population exerts pressure for more parks, playgrounds, sports complexes, hunting and fishing opportunities, hiking and jogging trails, and bikeways. Because of competing land uses, this is increasingly difficult to do. In many countries, hunting and fishing are no longer feasible options for the general public due to population density, declining game populations, and poor water quality. Some areas are too densely populated to set aside much land for local outdoor recreation.

Five of the most popular outdoor activities in the United States today are walking for exercise, swimming, bicycling, camping, and fishing. But even with just 75 people per square mile nationwide (unevenly distributed), recreational hunters and fishers complain about reduced local options. Hikers often need to drive a long distance to find a forest trail without "no trespassing" signs. More and more public and private lands are closed to the public, sometimes because of misuse by earlier visitors. Visits to federal recreation areas increased from 56.4 billion hours in 1980 to 80 billion in 1992, up 41 percent in 12 years, and state park and recreation areas hosted 725 million visits in 1992 (20).

Synthetic Materials

In the past few decades, the development of numerous synthetic materials has revolutionized the textile, container, utensil, packaging, rubber, and paint industries and greatly modified the products of many other firms. Among the most important synthetic materials are plastics, textile fibers, fiber glass, and rubber, which are rapidly replacing metal, wood, and natural rubber in many factory products. In the United States, consumption of paper and industrial minerals more than tripled and metals more than doubled between 1945 and 1973, but the rate of increase has since slowed appreciably due in large part to substitution of plastics and fiber glass for metals in many products.

Most of these synthetics are complex chemical compounds derived from organic materials, mostly plants, petroleum, and coal, but some are made from minerals. Plastics, like metals, can be heated and reshaped or poured into molds and will retain this shape when cooled. Celluloid, the first popular plastic, was developed over a century ago, but the modern age of plastics began in the 1940s and 1950s with nylon, vinyl, and various synthetic rubbers. Most are cheap to produce, durable, and relatively insoluble and stainless, so they are widely used in food and beverage containers, plastic bags and wraps, rainwear, tires, wood finishes, toys, films, tapes, lenses, dentures, artificial body parts, and other products.

Synthetic textile fibers, including nylon, orlon, dacron, rayon, polyester, and acrylic have largely replaced wool and linen, and compete with cotton in clothing, bedding, and other uses. Fiber glass is made from molten glass that has been forced through a sievelike device to obtain minute threads which are then recombined. It is used to make boats, bathtubs and showers, automobiles, and roofing.

Plastic's blessings are also their curse. Because plastic is cheap, it is frequently used just once and then discarded as litter and municipal waste. Because it is durable, it is difficult to incinerate and may lie for centuries in a landfill without disintegrating. To remedy this, a growing fraction of plastics are now being recycled and efforts are underway to develop cheap biodegradable (able to decay) plastics.

Economics of Resource Depletion

Some economists claim that the world will always have enough resources for any foreseeable population, because scarcities will increase prices, encourage the development of previously marginal resources, and hasten the invention and use of substitutes. In support of this, an engineer may rightly point out that we have the technology to provide housing and transportation for twice the present population, using high-rise apartment buildings and efficient urban transit systems. A county planner warns us that local jobs and tax revenues will be reduced if logging operations are curtailed to protect endangered species.

To counter these arguments, another economist points out that many of the projects we have the technology to do are not economically feasible. A wildlife biologist expresses concern about critical plant and animal species disappearing due to excessive logging. A hydrologist is alarmed because toxic substances are running into surface waters and seeping into undergroup aquifers. And other county planners wonder where they will get affordable land for future schools, parks, and low-rent housing for a growing population (Ch. 5).

The rapid expansion of mining, manufacturing, transportation systems, timber harvesting, commercial fishing, intensified agriculture, and even outdoor recreation imply increased pollution of air water, and soil unless effective remedial measures are taken. When they are not effective, serious consequences could result for human health, food supplies, the cost of living, natural ecosystems, and even climatic change.

World Resources Institute

One extensive source of current information on both natural resources and the human condition is the World Resources Institute, an independent, not-for-profit corporation that is supported by grants and donations (WRI, at 1709 New York Avenue, NW, Suite 700, Washington, DC 20006). WRI helps governments, the corporate sector, environmental groups, and others to find ways to meet human needs and foster necessary economic growth without destroying natural resources and environmental quality. WRI policy studies provide current information about global resources, environmental trends, emerging issues, and feasible responses to problems. WRI findings are shared through books, reports, papers, briefings, seminars, conferences, and media releases. Its biennial publication World Resources, prepared jointly with The United Nations Environment and Development Programme and the World Bank, provides up-to-date data on many of the topics addressed in this book. The 1996-97 edition features the urban environment, now the home almost half of the world's people, in collaboration with the 1996 United Nations' Habitat II Conference in Istanbul, Turkey.

Chapter 5

Environmental Degradation

We humans, especially since the advent of industrial technology, have radically altered our physical and biological environment and continue to cause changes at an increasing rate. The nature and importance of what is happening are perhaps best understood from the perspective of ecology, an interdisciplinary science that focuses on how living organisms relate to their environment and to each other.

An Ecological Perspective

The physical environment, our biosphere, is the layer of air, water, and land on the earth's surface that supports plant and animal life. It includes (1) continents and islands with their varied terrain, lakes and streams, accessible minerals, and groundwater, (2) oceans and seas that cover 71 percent of the earth's surface, and (3) the atmosphere, a mixture of gases. Many scientists view this biosphere as a thin, sensitive, delicately balanced refuge for life wedged between two hostile environments. And indeed, a few thousand feet below this life zone are intense heat and pressure, sometimes erupting to the surface as volcanoes or geysers. Just above it are the thin air, persistent winds, and perpetual cold of the lower stratosphere.

Not all scientists share this perspective of nature's vulnerability. Some say the biosphere is more hardy and resilient than commonly believed, noting that it has survived many severe onslaughts from nature itself, such as volcanic eruptions, hurricanes, floods, the ice ages, earthquakes, and collisions from meteors and asteroids. Each time it recovered from such disasters and when species were eliminated, others eventually filled their niche in the environment. Depending on the severity of the disruption and the variety of the species affected, recovery may take anywhere from a few months to many thousands of years.

Certainly life on the planet depends on the health and permanence of the earth's biosphere. The biosphere and all of the plants, animals, and microorganisms that share it constitute a giant ecosystem with millions of smaller but more or less integrated ecosystems. These subsystems range in size from a rotting log or a small marsh to a vast forest or an ocean. Each is a community of plants and animals that share a habitat, draw sustenance from it, and usually contribute something to it in return. Because the members of an ecosystem tend to be mutually supportive, impacts on one species may affect others and perhaps even disrupt the system. Usually each system is an integral part of a larger system, so further disruption could occur as a domino effect.

Sunlight provides the energy for plants to convert water and carbon dioxide from the air into carbohydrates (the process of photosynthesis), and to store it for future use. Birds, mammals, lizards, insects, and bacteria depend on the plants for food--and some of them eat each other. Other predators in turn (e.g., wolves, cats, eagles, or humans) eat them, a relationship called a food chain. Thus food chains extend "upward" from green plants and ocean plankton that depend on sunlight for their energy to the "highest" level, namely predators that depend on a series of species below them.

Each level upward on the food chain usually supports fewer organisms. This means that humans could be fed with fewer resources if they moved one notch down the food chain. Suppose, for example, that 12 tons of plant food were required to feed 10 hogs until they were mature enough to market. The same 12 tons of plant food could support more humans than eating the 10 hogs would, and also eliminate the need for the hog pen, pasture, and caring for the hogs. Hence, people in densely populated Asian countries eat more grain and less meat than Americans and Europeans.

Interruptions in the food chain occur when some species are forced to migrate or are exterminated. This affects other species that depend on them for food, as when the virtual extinction of the buffalo by in-migrant whites led to starvation for Plains Indian tribes. Pollutants (e.g., pesticides, carcinogens) absorbed by plants or animals at the lower end of the food chain may be passed on to those who eat them and increase in toxicity when they accumulate in their bodies.

In short, ecologists believe that a balanced, viable environment is essential to the survival and well-being of both humans and the numerous species that support us, yet we see mounting evidence that environmental quality is in jeopardy. We are wreaking havoc with the planet's ecosystems and the disruption of some of them affects others with adverse effects that are cumulative unless we take sensible measures to avoid or to reduce them.

Sources of Environmental Impacts

The combination of population growth, technological development, and the growing demand for natural resources have very diverse effects on the earth's physical and biological environment, many of them adverse and cumulative. Industries, agencies, institutions, farmers, tourists, and individual consumers everywhere generate or aggravate these conditions. All share the responsibility to recognize problems when they exist and to cooperate in solving them.

Frontier Philosophy of Abundance

Americans can learn a great deal from a review of their own natural resource practices since colonial times. Many of the industries that today depend on minerals, fossil fuels, timber, pastures, and croplands are haunted by this nation's earlier history of "reap, ruin, and run" development. Public policies encouraged rapid exploitation of natural resources nationwide. Various laws and executive decisions enabled farmers, miners, and railroads to obtain free land, timber, and minerals in return for accessing and developing natural resources.

Many people took what they wanted from the land without much concern for environmental quality and the resource needs of future generations. With a small national population and such bountiful resources, conservation was a difficult lesson to teach. People regarded the nation's natural wealth as virtually limitless and accepted little responsibility for the social and environmental consequences of their actions. Exploiting virtually free resources was in the spirit of the times and many who did became very wealthy.

Vast acreages of virgin forest were denuded by burning and clearcutting, resulting in flooding, soil erosion, falling water tables, and loss of wildlife. Forests were often logged without replanting and most grasslands were severely overgrazed. Strip mining devastated forests and croplands, while dredges and hydraulic mining raised havoc with streams. Mines and smelters desecrated hillsides, producing deadly air and water pollution, and contaminated sites were abandoned without reclamation. Bison were killed by the thousands for their hides and tongues, or shot from train windows for sport. Birds were slaughtered by the hundreds for feathers or diversion and left to rot. Many types of sea creatures were (and still are) needlessly killed along with marketable fish.

Subsequent writers often call the early captains of industry "robber barons," because they took so much from the environment and gave so little in return. Their ranks include many famous names in history, people who harvested the country's resources on a massive scale, abandoned exploited areas for virgin areas, and established family fortunes. But small businesses and individual families also played a role in reducing the nation's immense store of resources through wasteful use. With today's technology and increased needs, much of what was wantonly killed, burned, wasted, or cast aside would now be regarded as a valuable resource. We cannot reclaim most of these misused resources but we can learn from the past mistakes.

Industrial Activities Today

The conquest of nature continues on an ever-larger scale, even though more restrictions have been imposed to reduce deliberate waste and degradation of the planet's resources. Mammoth earth moving machines dig canals, level hills, drain swamps, fill coastal wetlands, and clear the landscape for buildings and freeways. High dams reduce or prevent the age-old migration of salmon and other anadromous fish, while the reservoirs behind them retard the flow of streams, collect silt, and sometimes interrupt the seasonal migration of big game.

Thousands of great factory complexes and hundreds of millions of vehicles and engines spew billions of tons of noxious gases and particles into the planet's atmosphere. Giant open-pit and shaft mines and deep oil wells intersect and pollute underground water sources and sometimes produce slumps on the surface. Pesticides, fertilizers, industrial chemicals, and domestic animal wastes are absorbed in groundwater, flow into streams and lakes, contaminate drinking water, and kill vegetation and fish (1).

Key Areas of Concern

Major areas of environmental concern today include air and water pollution, global warming, the thinning of the ozone layer, massive extinction of plant and animal species, tropical deforestation, soil erosion and depletion, the extension of deserts, the release of toxic chemicals, and the creation of mountains of waste. The following sections explore each of these conditions.

Mining and Drilling

Total world minerals consumption is roughly doubling each generation (25-30) years, faster than the world's population, which doubled in the past 40 years. Large areas are still being laid to waste by strip, hydraulic, and dredge mining, and by toxic fumes from smelters. Some mined areas have remained unproductive because disturbed terrain is too uneven, infertile, or toxic to support vegetation without extensive reclamation.

A review of present reclamation needs suggests the immensity of the challenges facing future generations. Each year the world's mining industry produces billions of tons of refined minerals materials. To produce this volume of commodities, usually much more earth is removed to access the ore and to separate the desired minerals from other materials. The waste removed in mining each commodity averages less than 50 percent for easily accessed, naturally concentrated substances (such as stone, clays, sand, or gravel), but is about 60 percent for iron, 77 for aluminum ore, 91 for phosphate, 99 for copper, and much higher for some precious and rare metals and gems (2). It is estimated that an area roughly the size of Rhode Island is disturbed each year by mining operations worldwide, excluding oil and gas. Direct effects to the land include removal of overburden, separating and depositing unwanted materials from the ore (tailings), constructing processing plants, creating transportation and utilities corridors, and cutting timber for tunnel supports. Other common impacts are displacing human and wildlife populations, water and wind erosion, and releasing toxic pollutants that affect air, groundwater, and soils. In the extreme case, fumes from a single smelter can kill thousands of acres of vegetation and decimate fish populations in lakes a hundred or more miles downwind.

These multiple effects demonstrate the importance of careful mine and ore processing plant design to minimize environmental impacts. Competent corporate and agency supervision of mining and minerals processing operations is needed to ensure compliance with environmental laws and effective reclamation of abandoned sites. The land base for developing new ore deposits is now shrinking due to earlier exploitation of the richest deposits, competing land uses, and more rigorous environmental and health restrictions on mining activities.

Petroleum development also poses environmental risks. Oil wells, sometimes over a mile deep, cut through many layers of rock, some yielding water. Increasingly, drilling crews are well equipped and able to minimize contamination of water bearing strata. But potential problems exist, including drilling into poisonous gases or salt water, pipeline leaks, fires, and illegal dumping of wastes. Related activities, such as noisy seismic testing, road and pipeline construction, transporting oil and gas, and rapid changes in personnel and logistical support, affect both the natural and the social environment. Although the petroleum industry has done much to reduce or to eliminate these problems, substantial risks remain, especially in areas where environmental regulations are weak or ineffective and infrastructures are relatively primitive.

Soil Erosion and Depletion

Even though material living standards are rising in many countries, they are declining in others that lack the resources and technology to meet the needs of their people. According to the United Nations Environment Program, 11 percent of the planet's soils, some three billion acres, have been severely degraded in the past 45 years. This is an area considerably larger than either Europe or the United States. Vegetative cover and soil fertility are depleted by decades of harvesting native trees and other plants, insufficient replanting, overgrazing, and sowing the same crop repeatedly, especially on dry or infertile soils. Massive wind and water erosion sometimes follow, forcing farmers to revitalize their soils or to move to other areas.

The severe dust storms on the central plains of the United States in the 1930s occurred after the removal of most of the native grass and shrubs to permit seasonal grain farming. Years of drought followed, during which hundreds of millions of tons of topsoil were blown away from farmlands in several states. Many farmers went bankrupt and moved elsewhere, a tragedy vividly depicted in John Steinbeck's 1939 novel, *Grapes of Wrath*.

Soils also become less productive when they accumulate salt residues from irrigation or sea waters, absorb toxic chemicals from natural or industrial sources, or become excessively acidic. Compaction from vehicles and livestock can interfere with drainage, inhibit plant root development, and reduce yields. A farmer with the necessary capital and expertise can enrich depleted soils by testing them, adding essential nutrients, and irrigating or fallowing them between plantings. Erosion can be reduced by strip farming on level fields, contour farming and terracing on hillsides, planting shelter belts, and installing slow-flowing irrigation systems.

Desertification

Desertification refers to increasing dryness and reduced productivity on croplands, usually areas located on the margins of deserts or arid grasslands. This transition has been evident for years on the margins of the Sahara Desert, in central Asia, and parts of southwestern U.S. To some extent desertification is due to natural climatic cycles, but human activities also play a role.

As human populations increase and extend into drier areas, native grasses and shrubs are cut for firewood or to clear additional fields. Plowing the thin, fragile topsoil and overgrazing by livestock increase the potential for rainwater runoff and wind erosion. Local fresh water consumption, crop irrigation, and the loss of rainwater lower the water table. When a period of drought intensifies the situation, some people are forced to transport water long distances or to migrate elsewhere.

Deforestation

Centuries ago about a third of the earth's land area was forested and some foresters claim a fourth of it is still predominately forest land, even though less than half of what remains is similar to its natural state with relatively undisturbed ecosystems. Through the generations, much of the world's original temperate forest cover has been cut for timber, fuel, and wood pulp, or burned to clear land for crops, pastures, and building sites (Ch. 4). In many countries, portions of the remaining virgin forests have been set aside as national forests or parks.

Only about 40 percent of Europe's earlier forest cover remains and almost none is natural, random-aged growth. Most of these forested areas and many in the eastern U.S. are scattered rural groves and second-growth public and commercial forests. The lush old-growth forests of the U.S. Pacific Northwest remained fairly intact until World War II but since then both federal and private lands have been extensively harvested. Larger areas of relatively unbroken forest are found in Canada, Alaska, and northern Russia, but sections of these too are falling to the chain saw.

National parks and forests around the world are also under siege by poachers, timber and minerals rustlers, and unauthorized farming (both conventional and illegal crops, such as marijuana). Other threats include pollution from heavy traffic, adjacent development, smoke, airborne chemicals, and alien pests and diseases, such as the gypsy moth and Dutch elm disease.

Timber harvesting methods are frequently criticized because of their environmental impacts, including networks of roads in steep terrain, clearcutting, smoke pollution, and adding sediments to streams that affect water quality and fisheries. Clearcutting is the practice of harvesting all of the trees on large tracts and leaving gaping holes in the forest that are littered with stumps, clumps of brush, and scattered debris. This technique is the most cost-effective way to harvest dense forests and has been widely used in the Pacific Northwest, Canada, Siberia, and the tropics.

Clearcutting's defenders say that landowners should be able to harvest as they see fit, that it creates meadows for big game, and that the areas quickly recover. Critics of the practice argue that it disrupts ecosystems, destroys wildlife habitat, creates eyesores in place of natural beauty, and greatly increases erosion from heavy rains or floods. They advocate reduced roading, selective cutting of timber, and leaving the forest relatively intact with vegetation of different ages and healthy ecosystems.

Extensive tropical hardwood forests have survived into the modern era, including large segments of the vast evergreen expanses found in Central and South America, central Africa, southeast Asia, the Philippines, and Indonesia. For many centuries, some degree of deforestation by humans occurred in the tropics, but never at the present rate. Slash and burn agriculture was a common practice of forest residents who cleared small plots of land by burning and then planted crops. When the fragile soil was depleted, they simply abandoned their gardens and prepared new ones. Local population was kept in check by natural causes (Ch. 2), the plots were scattered, and the abundant rainfall and year-round growing season encouraged the growth of new vegetation.

The situation is very different in many areas of the tropics today because of rapid population growth, increased trade with other countries, and extensive exploitation of natural resources by foreign as well as domestic firms. In addition to the traditional slash and burn agriculture, land is being cleared on a massive scale for farms and plantations, roads, timber exports, mines, oil fields, hydroelectric power development, housing, and utilities corridors.

The rate of tropical deforestation increased 60 to 90 percent during the 1980s and by 1991, about two percent of the forested area was being eliminated each year. The loss is especially great in southeast Asia, Madagascar, West Africa, Central America, and the eastern Amazon region, where some experts fear that most forests will be severely depleted within two decades if harvesting continues at present rates (4). In Madagascar, almost two-thirds of the forest is now removed to provide cropland for a growing population and more is cut each year, greatly reducing the habitat available to its many unique species.

Tropical rain forests are extremely valuable to their countries and to the world. Their persistent warmth and humidity provide an ideal environment for a great diversity of plant and animal species, estimated at half or more of the world's total. Cutting and burning reduces or eliminates the population of many of these species with potentially adverse consequences for humans and wildlife. Both local and multinational industries need these forests for renewable raw materials used in foods, drugs, drinks, and various other products. Plant geneticists depend on them for both the ancient and newer varieties of plants needed to improve or to replace existing domestic species. Trees store great quantities of carbon and releasing it through burning increases carbon dioxide in the atmosphere, hastens global warming, and apparently alters climatic cycles. As in temperate zone forests, reduced vegetative cover in the tropics results in increased soil erosion, more lake and stream sedimentation, and more frequent flooding. Foreign tourism, an important source of stable currencies for developing nations, declines in areas with desecrated landscapes and persistent smoke.

Finally, tropical forests are the home of many people, some of them nonliterate tribes who have lived there for centuries and have limited contact with the outside world. These and other local inhabitants traditionally harvest renewable forest products, such as fruits, nuts, roots, fuel, fiber, and thatching materials for their own use, for local sale or barter, and for export. Their way of life threatened as outsiders cut their timber and convert their forests to other uses.

Total deforestation need not be inevitable, but stern measures are needed to reduce the rate of increase in exploitation. Both foreign and domestic industries are reaping sizable profits by depleting forests and are not held responsible for much of the social and environmental devastation that follows (Ch. 7, 10). More countries are now taking steps to protect their remaining virgin forests, some are replanting abandoned farm and mine acreage, and the rising cost of lumber encourages the use of other materials.

Air Pollution

Fresh air is a mixture of 78 percent nitrogen, 21 percent oxygen, and 1 percent other gases, mainly argon and carbon dioxide. "Natural" air also contains varying amounts of water vapor (or ice crystals), dust, pollens, salt particles, and assorted microorganisms. Humans and other animals breathe the oxygen, whereas plants absorb carbon dioxide and release oxygen.

Population growth, expanding industries, modern farming methods, and related infrastructure development have increased the type, scale, and seriousness of environmental pollution. Some types of air and water pollution are now global in scope because prevailing winds and ocean currents circulate worldwide. Fortunately we are also seeing greater international vigilance and cooperation in efforts to control pollution (Ch. 8, 11). Air pollution usually refers to air contaminants resulting from human activities, particularly when they jeopardize the health of humans, wildlife, and plants. Pollutants include

--solid particles (called particulates), such as coal and asbestos dust, soot, and lead,

--liquid droplets, including insecticides and herbicides,

--toxic gases, such as carbon monoxide, nitrogen dioxide, and sulfur dioxide,

--radioactive fallout, and

--combinations of these, as from power generation, smelters, factory emissions, vehicle exhaust, and tobacco smoke.

World carbon emissions, which produce carbon dioxide, the major cause of global warming, reached six billion tons in 1995. Industrial nations contributed about half of this total, led by the U.S. with 1.37 billion tons. Industrial operations account for about 40 percent of these emissions, vehicles for 20 percent, households for 20 percent, and the remainder stems from other sources including agriculture and forestry (5). Petroleum fueled engines (e.g., in cars, trucks, trains, jets, pumps, chain saws, and lawnmowers) emit a combination of noxious gases, including carbon monoxide, nitrogen oxides, hydrocarbons, and particulates (chiefly lead). In one recent year, the United States, with about 25 percent of the world's vehicles, released an estimated eight million tons of nitrogen oxides and six million tons of hydrocarbons into the atmosphere. Coal and oil are burned to provide heat in homes, offices, and industries, to generate electrical and steam power, to process industrial materials, and to incinerate garbage. These uses emit the same gases as engines plus toxic sulfur dioxide. Gaseous pollutants combine with water vapor in the air to form visible smog that can linger in an area for days when the wind is still, endangering plants and animals as well as humans. Smog is a factor in numerous deaths (4000 in London in 1952), aggravating preexisting conditions such as asthma and bronchitis, and shortening the lives of residents in smog-ridden cities. These gases also reduce human efficiency, induce headaches and other illnesses, and may contribute to birth defects. Acid rain occurs when high levels of sulfuric and nitric acid combine with water vapor in fog, rain, or sleet. This mixture sometimes travels hundreds of miles downwind and can contaminate water, damage vegetation, kill fish, and erode buildings and other structures.

The collapse of the Iron Curtain revealed to the West a large area of serious industrial pollution extending from Magdeburg, in former East Germany, to Dimitrovgrad, Bulgaria, 900 miles south. This is a region of extensive industry with outdated mines, smelters, and factories, and ineffective pollution controls. As a consequence, over a million acres of woodlands in Poland and Czechoslovakia were damaged by smoke and acid rain. In the most blighted areas, flocks of white sheep were black from soot and many people suffered from respiratory disorders (6). Visitors to the area were astonished by the level and extent of pollution, which local people resented but could not remedy. The United States and western Europe had similar problems during earlier years of industrialization, but federal, state, and local efforts have been increasingly effective in reducing air pollution (Ch. 8).

Global Warming

Each ton of carbon released into the air by human activities mixes with air to produce 3.7 tons of carbon dioxide, which accumulates and acts as a barrier to prevent some of the sun's heat from rising into the upper atmosphere. The World Resources Institute estimates that the people of the world added 26.4 billion metric tons of carbon dioxide to the atmosphere in 1992, with 84 percent due to industrial activities (broadly defined) and the U.S. contributing almost 22 percent (7). Large-scale burning of the world's tropical forests adds both carbon and particulates. Other gases, such as methane, nitrous oxide, and chlorofluorocarbons (CFCs), also block heat radiation from the earth. This is the greenhouse effect that growing numbers of scientists believe is the principal cause of global warming.

U.S. scientists tracking the global warming trend have determined that 1990 and 1995 were the hottest years on record, the average global temperature has risen one degree Fahrenheit this century, and the increase appears to be accelerating (8). The latest (1995) United Nations report on climate change concurs, noting that the 10 warmest years since 1860 have occurred in the past 15 years (9). This warming is uneven, with higher latitudes generally more affected than the tropics. In 1991 the eruption of Mt. Pinatubo in the Philippines injected enough dust into the atmosphere to reduce warming temporarily but this has now settled. Fluctuations in the sun's intensity also affect global temperatures, but this factor may now be eclipsed by human activities.

During this period, mountain glaciers have been receding in many areas, the sea level rose several inches due largely to heat-induced expansion, and parts of Antarctica's ice shelves have begun to break up with large segments floating away, eventually to melt. Other evidence of global warming is the increase in unstable weather in the earth's temperate zones, including unusually heavy rains and the 1995 heat wave that killed over 500 Chicagoans.

How much additional global warming will occur depends on other factors as well. Both long and short-term climatic cycles are normal, so the accumulating carbon could intensify a warming phase in the cycle or help to neutralize a cooling phase. The Environmental Protection Agency (EPA) projects that global emissions from human activities could double by 2025, with developing nations contributing most of the increase, and this could accelerate the warming trend. Humans might also be doing other things to the environment to accelerate or reduce a warming trend.

If global warming continues long enough, the deep ice caps of Antarctica and Greenland will gradually melt and eventually release enough water to inundate lower portions of heavily populated coastal plains around the world, displacing up to two billion people. Tropical storms could be more frequent and violent, estuaries and coastal wetlands would be flooded, and many plant and animal species could become extinct if they failed to migrate to higher ground or to cooler habitats. Increased warmth could enable insect pests and harmful microorganisms to extend their range. Global warming could make also deserts and grasslands even drier, while extending croplands farther north in Canada, Siberia, and Alaska. In any case, global warming is a slow trend giving nations time to retard it by various means and to evacuate threatened areas before they are flooded or excessively arid.

Because the United States consumes so much commercial energy, its contribution to global warming is far greater than its five percent of the population would suggest. This level of energy consumption exceeds the entire continent of Africa and may be as great as China and India combined, with their two billion people. As these developing nations industrialize, motorize, and grow in population, their contribution to global warming is expected to increase much faster than in the West.

Ozone Depletion

Ozone is the most chemically active form of oxygen, created on the earth's surface when electrical discharges arc through the air and when nitrogen oxide and hydrocarbon pollutants interact. In higher concentrations, as in urban smog, it is unhealthy to breathe and can harm plants as well.

A more beneficial ozone is created when the sun's ultraviolet rays interact with oxygen in the thin air that surrounds the globe about 20 or 25 miles above its surface. This stratospheric ozone layer acts as a barrier to ultraviolet radiation and protects humans and other organisms from excessive exposure. But the release of chlorofluorocarbons (CFCs, such as freon) from refrigerators, air conditioners, and spray cans weakens this protective layer because chlorine in them drifts upward and destroys ozone molecules. As a result people become more susceptible to sunburn, skin cancer, and eye disorders.

In 1985 researchers discovered a growing hole in the ozone layer over Antarctica and deduced that the intense cold enhances this destructive process. To counteract this threat, most of the nations of the world are phasing out the production of CFCs by the year 2000. Ultimately ozone levels may again increase but it will take decades for these gases to dissipate.

Water Pollution

Industries such as mining, minerals processing, manufacturing, and timber are responsible for much of the water pollution in industrial countries and for the release of many of the most toxic chemicals. They use large quantities of fresh water to convey and process wastes, some of which contaminate local groundwater or are discharged into rivers, lakes, or the ocean. Common industrial wastes include lead, mercury, sodium, caustic acids, asbestos, petrochemicals, cement, gypsum, abrasives, solvents, gases, and PCB (a hydrocarbon containing chlorine). Farms and ranches use large quantities of nitrates and phosphates as fertilizers, and insecticides and herbicides to control pests. Some of them have densely populated livestock feedlots and poultry barns. Agricultural chemicals and animal wastes seep into the ground or are washed along with eroded soils into streams and rivers by heavy rains, adversely affecting wells and fisheries. Some countries still use DDT, a potent insecticide that is harmful to birds, fish, and animals, including humans. It was widely used in the United States in the 1950s and 1960s but was banned in 1972 due to increased concern about its toxic effects.

When dams are constructed to store water, control floods, and generate electrical power they impede the normal flow of water, limit fish migration, and encourage sedimentation. When runoff fertilizer enters the stream, some native plants and marine life are likely to die out while other plants grow profusely, a condition called eutrophication.

Municipalities and federal and state installations are a third source of pollution. Some release partially treated sewage into local streams, especially when rainstorms overload their systems. Most solid wastes and bacteria are removed in the sewage treatment process, but viruses, tastes, and odors often survive. Service stations sometimes have leaky tanks or spills and release gasoline into porous earth. Do-it-yourself car owners, shop employees, and homeowners sometimes pour their oil, paints, solvents, and other toxic substances into their house or shop drains or on the ground and they seep into subsurface water.

The Ganges River in India well illustrates how bad river pollution can get. Each year 29 cities and thousands of towns and villages release 1.3 billion liters of wastes into the river. Assorted industries add a fifth more and some 340 billion tons of sediment are picked up from natural erosion, forest clearing, and construction activities. Add to this 6 million tons of chemical fertilizers, 9000 tons of pesticides, and the ashes of thousands of cremated Hindus. Currently this water is used for bathing, cleaning, and drinking, and in another generation a billion people will share this river basin (10).

Toxic chemicals and wastes have known effects on human and animal health. Depending on the type and dosage, they may cause lung, liver, or nerve damage and are also suspected as the cause of some cancers and birth defects. Other wastes may cause nausea, diarrhea, hepatitis, salmonella, or allergic reactions and render fish and shellfish unfit to eat. Medical wastes are sometimes improperly handled, allowing the spread of diseases.

Ocean Pollution

Oceans and connected seas cover 71 percent of the earth's surface and almost two-thirds of the world's people live within a hundred miles of their shores. Ocean health is vital to its indigenous plants and animals, to humans and wildlife that depend on ocean species for food, and to the quality of the air we breathe. These waters support many valuable species, beginning with plankton, the minute plant and animal organisms at the bottom of the food chain that drift near the ocean's surface. Plankton and a variety of larger plant life consume carbon dioxide and release oxygen to the atmosphere, while ocean currents and relatively constant ocean temperatures help to stabilize climates.

Other widespread ocean species include enormous colonies of coral, squid, octopi, jellyfish, giant turtles, thousands of species of fish and shellfish, numerous aquatic and shore birds, and diverse mammals, such as whales, dolphins, seals, walruses, and sea otters. All of these participate in intricate ocean ecosystems and the demise of some of them from overfishing, pollution, or other causes affects others.

Oceans are the final destination of water from most of the earth's rivers, lakes, and streams, so they accumulate water-borne pollution, both chemicals and vast amounts of solid wastes. They are further contaminated by oil spills from tankers and drilling rigs, by coastal industries, and by the now illegal practice of dumping municipal garbage and refuse from ships into the sea. Dumping at sea was prohibited by international agreement in 1988 but it is hard to enforce.

Chemical pollutants, solid wastes, and physical contact can kill or injure plant and animal species, including colonies of coral, especially along the shallow shelves surrounding the continents and heavily populated islands. Aquatic mammals and fish sometimes die from eating plastics, metals, and other indigestible solid wastes, or from entanglement in nets and other materials. Thousands of birds, fish, and mammals are exterminated in oil spills, such as the 1989 Prince William Sound disaster in Alaska and the 1996 spill on the New England coast. Swimming in some coastal areas is now prohibited due to pollution and one can readily imagine the effects of this water on fish and plants harvested for human consumption.

Coral and aquatic plants are also killed by underwater explosions and by abrasions from boats, nets, and other equipment. In addition, the thinning of the ozone layer allows harmful rays of the sun to kill plankton, thus reducing the supply of food for many species along with the ocean's ability to stabilize atmospheric gases by consuming carbon dioxide.

Other Sources of Pollution

Noise pollution refers to sounds that are loud and persistent enough to damage the hearing or cause emotional stress. Typical sources in modern life are construction projects, jet landing strips, heavy urban traffic, boomboxes in cars and on city streets, and loud television or stereo sets that are audible through poorly insulated walls of apartments or motel rooms. Eye pollution is another vexing problem in a society that values scenic beauty and is proud of its high standard of living. Some offensive sights are more common in the United States than many western European countries, including huge unfenced junkyards near public highways or institutions, abandoned buildings left to deteriorate, homes with neglected yards made conspicuous by tall weeds and assorted refuse, dozens of huge billboards near each city, fast food litter along streets and highways, and wastes dumped on parking lots, often just steps away from a trash can.

Species Extinction

Scientists have identified at least five massive extinctions during the long period of life on earth. The most recent one, now attributed to an asteroid falling into the Gulf of Mexico about 65 million years ago, disrupted the earth's ecosystem and caused the extinction of the dinosaurs along with numerous other species. Less dramatic acts of nature, such as gradual climatic changes, volcanic eruptions, floods, and wildfires also reduce the population of various species but are unlikely to kill all of their members, so most of them gradually recover. When extinctions do occur, other species often increase or otherwise adapt to restore balance in the ecosystem.

Now biologists are warning us of a sixth major extinction, one caused by humans and already well underway. A recent United Nations report, "Global Biodiversity Assessment," is said to be the first worldwide assessment of Earth's living resources. It estimates that the planet is inhabited by as many as 14 million living things, more than many earlier estimates, although only about an eighth have been cataloged and described. The report states that species are now disappearing at about 1,000 times the usual rate, principally due to human activities (11).

During the past 400 years, more than 1100 plant and animal species are recorded as becoming extinct. But many thousands of other known and unknown species, including over a hundred kinds of birds and mammals have also become extinct. Additional tens of thousands of species are now committed to extinction, and even more thousands of yet-unnamed organisms will continue to die out each year. Harvard biologist Edward Wilson estimates that at least 50,000 species are doomed each year by the destruction of tropical rain forests (12).

To cite some examples, worldwide declines are now evident in three-fourths of bird species, among turtles and frogs, in numerous species of fish, shellfish, and ocean mammals, and among various land mammals, including primates, bears, wild cats, elephants, wolves, and many others. One source estimates that Kenya's elephants have declined from 169,000 in 1969 to about 20,000 in 1990 and that black rhinoceroses, killed for their horns, have declined from 65,000 in 1970 to 4000 in 1990 (13).

Overfishing, wasteful fishing practices, and ocean pollution have greatly reduced the population of many ocean food species, such the large whales,

Atlantic cod, haddock, red snapper, flounder, anchovies, and lobster. Trawlers with large nets catch and suffocate or discard many unintended species, including porpoises, dolphins, turtles, and birds. Coral is damaged by heavy nets and by scuba divers collecting souvenirs.

Historically in the United States, hundreds of familiar species that took millions of years to evolve were either decimated or became extinct because of unregulated hunting, trapping, habitat conversion, waste dumping, toxic substance spills, pesticide applications, and other land-use practices. Well-known examples are the bison, caribou, wolves, mountain goats, whooping cranes, bald eagles, peregrine falcons, passenger pigeons, sea otters, seals, and many medically valuable plants (Ch. 4). However, whooping cranes, bald eagles, peregrine falcons, sea otters, and bison are now increasing in numbers because of recent efforts to protect them.

One very noticeable loss in both urban and rural areas is songbirds. Many species have fallen victim to habitat destruction, lawn and field pesticides, human consumption, and the mushrooming population of free-roaming house cats. Aside from the beauty of their songs and their captivating antics, birds are important in controlling insect pests and pollinating plants. A study in Wisconsin estimated that 19 million songbirds and additional game birds may be killed annually by domestic cats (14). Studies in the State of Victoria in Australia and in Great Britain yielded similar findings. The Humane Society estimates that cat and dog populations in the United States have grown to over 50 million each, that 10 million pets are unwanted and euthanized annually, and that over 3000 animal shelters are filled to capacity.

Wildlife biologists point out that a great diversity of species in an area contributes to the stability of an ecosystem. The extinction of one species in a complex system usually causes fewer problems than in a simple system and it is less likely that one species will have runaway population growth or harm the environment. Some creatures, such as the spotted owl or certain frogs, are called "indicator species" because the state of their health and the number still surviving are regarded as indications of the general health of forests or wetlands where they dwell.

Another cause of species extinction is the accidental or intentional introduction of alien species that attack and kill or weaken native plants or animals, or simply grow profusely and displace other species. Exotic species often thrive in the absence of natural enemies in a new ecosystem. The list of examples is long, including the gypsy moth, starling, bee mite, and Dutch elm disease from Europe to America, the rabbit from Europe to Australia, Japanese beetles to the United States, the potato beetle and blight

76   *The Environmental Dilemma: Optimism or Despair?*

from Mexico to Europe and the United States, the so-called killer bee from South Africa to the Americas, and AIDs (acquired immunodeficiency syndrome) from Africa to the globe. The USDA estimates the 1981 loss from the gypsy moth alone at $764 million. The bee mite has already exterminated much of the U.S. wild bee population and poses a serious threat to domestic honey production. The loss from AIDs is incalculable.

Programs are now underway in a growing number of countries to protect endangered species by securing their habitat and restricting hunting, fishing, and polluting of fisheries. This effort is hampered by poaching and the illicit international traffic in tropical mammals, birds, other wildlife, rare pelts, horns, and ivory. Smugglers violate laws regulating the import and export of such items and pose the added risk of transporting destructive insects and infectious disease to other countries.

Solid Waste Disposal

Most of the raw materials so essential to industrial economies eventually become wastes and often counties and cities, rather than the producer or the manufacturer, face the problem of disposal. Like consumption, solid waste has been increasing because each year there are more people, each person consumes more goods, and often packaging is more elaborate (Ch. 3).

Each American disposes of close to a ton and a half of garbage annually and this output is increasing, even though a growing portion of it is now being recycled. Much of this refuse consists of paper products, plastics, food wastes, aluminum, other metals, discarded furniture, and lawn and garden debris. Each year Americans alone (five percent of the world's people) dispose of literally tens of billions of cans, glass bottles, and other paper and plastic containers, plus countless tons of disposable diapers, paper towels, plastic toys and gadgets, and slick magazines. Much of this is used only once, is not recycled, and becomes part of our mountains of solid wastes.   In addition to household and office garbage we dispose of municipal sewage, mining and industrial wastes, millions of discarded cars and appliances, and enormous amounts of material from demolished buildings, highways, and sidewalks. Much of our solid waste is virtually indestructible and accumulates from one generation to the next unless it is recycled or reused.   Over 70 percent of U.S. solid garbage is deposited in landfills and the remainder is burned or recycled. Landfills displace other land uses, potentially contaminate underground water supplies, make recyclable materials inaccessible, and can be an eyesore. Critics of U.S. waste disposal policies note that Germany, Japan and several other industrial nations produce half as much waste per capita as Canada and the United States, and burn or recycle more of it (15).

Nuclear Wastes

Scientists have not yet found a safe, permanent way to dispose of all of the waste from commercial atomic energy plants and other sources. Most of this waste is a by-product of electric power generation, which currently provides about five percent of the world's supply. In 1994, the U.S., with 109 of the world's 421 nuclear reactors, had accumulated over 29,000 metric tons of this waste (16). Only time, over 100,000 years for some wastes, reduces the harmful radioactivity. Burying these materials in the earth does not ensure that they will never leak radioactivity to the surface.

Taxing the Earth

Until about 1900, public criticism of unrestricted population growth, unfettered natural resource development, and needless environmental damage was too weak and fragmented to encourage much legislative and regulatory response. Today, in the face of growing public concern and increased federal and state involvement, some conscientious firms and municipalities strive to meet consumer needs while avoiding or minimizing environmental impacts. Many others are unlikely to conform unless agencies set standards and supervise their operations.

Nature itself is at times very destructive to the physical and biological environment, as demonstrated in numerous examples above. But the earth also has remarkable powers to heal its wounds. Rain cleanses the air of pollutants, bacteria consume plant wastes and convert them into nutrients, new growth quickly replaces burned or windblown areas, birds eat insect pests, and new species in-migrate or slowly evolve to fill niches in incomplete ecosystems.

The reality that humans must face is that their own pollution and devastation is in addition to natural processes and sometimes overwhelms nature's ability to repair itself, at least in the short term. We have noted that garden pesticides and house pets kill the birds that would help to control insect pests. Often we respond by using more pesticides and killing more birds. One can also cite instances of human intervention to restore the environment; for example, efforts to protect birds, to stabilize an ecosystem, or to reduce wind and water erosion. These offset some of our depredations but are often insufficient to restore the balance.

It is essential that the earth retain its ability to meet both today's needs and those of hundreds of succeeding generations of humans. To do this, we must ensure that ecosystems and other natural systems vital to our existence, directly or indirectly, also remain viable. This is a formidable challenge but the nations of the world are beginning to respond, sometimes with impressive results.

Chapter 6

Social Effects of Industrial Development

The industrial revolution that began in Europe and North America over two centuries ago extended over several generations but many changes were unprecedented and their long-term effects were difficult to foresee. Equally sweeping changes sometimes occur much faster in the developing world today, but these nations have the advantage of hindsight and can avoid some pitfalls by learning from others when feasible.

Each developing nation must endure a period of social and economic upheaval as shifts from rural subsistence agriculture to industrial production in urban settings (Ch. 1). A sequence of rapid population growth, limited croplands, rural unemployment, natural resource development, expanding industry, and heightened awareness of new opportunities forces the relocation and cultural adaptation of millions of rural people. The transition induces profound changes in technology, social organization, traditional outlook, employment options, living arrangements, and consumption patterns.

Global Scale of Economic Growth

Economic expansion is now occurring at millions of sites around the world. It takes many forms: new or expanded mines, industrial plants, roads and utilities, housing, wholesale and retail sales outlets, and commercial services. Economic growth has literally millions of sponsors, including large corporations, the World Bank, small businesses, private aid programs, and individual investors. Federal, state, and local governments also establish and operate commercial enterprises, such as utilities, farms, and shops, and contract for goods and services. Industrialization and subsequent modernization in each nation contribute to the environmental dilemma by intensifying population growth (at least temporarily), utilizing large quantities of natural resources, and degrading the natural environment (Ch. 2-5). Ironically, this transition also increases each nation's potential to curb population growth and to protect and restore the environment if it so chooses and gradually more of them are doing these things.

The combined social impacts of so many independent economic efforts are difficult to predict or to control. Depending on who is affected and how, a given innovation will delight some people and disappoint others. Consider, for example, both the social and the health effects of the emergence of family cars and public transportation in the U.S. Think of the many related changes, such as the creation of suburbs, fast foods, drive-in restaurants, cars for teenagers, traffic accidents, noise and exhaust, and crowded national parks.

We can welcome the convenience and appreciate the growing popularity of newer products designed for disposal after one use, such as containers, diapers, razors, tableware, and even cameras. But we must also consider their implications for resource depletion, litter, and solid waste disposal. Yet these goods and services have become so popular that in some locations alternatives are no longer available.

Charting the Course

Since the 1930s, scientists and agency personnel have monitored some key social and environmental conditions, thus increasing our understanding of what is happening and why, and improving our ability to forecast trends. The *Statistical Abstract of the United States*, revised annually, provides hundreds of examples of such data. With experience, we are becoming more adept at foreseeing problems and suggesting feasible solutions. However, funding for scientific research is both limited and uneven, favoring some areas and neglecting others of equal or greater importance in the long term.

Public officials who keep abreast of these trends may be aware of impending problems but still lack adequate information for formulating workable solutions. They also have limited resources and are sometimes subject to intense political pressures. Some of them find it more expedient to heed special interests or their superiors (with reelection or promotion in mind) than to do what they think is right (Ch. 10). In the U.S., efforts are underway to reduce the influence of special interests by limiting the sources and size of political campaign contributions. Other nations try to do this by establishing quasi-independent agencies that monitor conditions, formulate national goals, and plan for growth and change.

Costs and Benefits of Development

Many types of industrial development are widely popular and seem to be inevitable in one place or another. The pressure of population growth, the quest for profits, and the pursuit of higher material living standards are no doubt strong motives for economic growth (Ch 1-3). But such growth is seldom a "win-win" situation because it generates costs as well at benefits, both at the local (micro) level where the direct effects are most apparent and for the larger society (macro level) where at least some indirect effects are evident.

Profit as the Primary Motive

Usually the primary goal of a private business or industry is to enhance the welfare of its owners, managers, and investors by increasing its sales and earning a profit. Unless it at least recoups its costs, it may be forced to close down. The quest for greater profits is an incentive to contain costs by eliminating inefficient operations, paying low wages to most employees, trimming the work force, and paying bottom dollar for supplies and equipment. Secondary goals vary with the firm and may include filling a vital need (e.g., essential food or medicine), producing a quality product (e.g., a world-class auto), providing safe, pleasant working conditions, or avoiding environmental damage.

To illustrate the strength of the profit motive, consider past decisions of U.S. manufacturing firms to close hundreds of plants (e.g., textiles, electronics, steel, and shoes) and to move these operations abroad to take advantage of much cheaper labor and fewer environmental restrictions. Corporate managers knew very well the stress and turmoil that would follow when dedicated employees find themselves jobless, lose pension and health benefits, and cannot meet family obligations. "Tough" managers were amply rewarded for making such decisions and stockholders with numerous and revolving investments were often only dimly aware of what was happening within these firms. Once some firms made theses changes to reduce their costs, others felt pressured to do the same to remain competitive.

Recently many corporations have downsized their domestic operations and are increasing profits through massive layoffs, an estimated 400,000 cutbacks in 1995 alone. Here too, the result has been the agony of joblessness for displaced workers, the frantic search for new jobs (sometimes at half the salary), increased reliance on unemployment insurance, and a greater demand for public social and medical services. These millions of layoffs decreased the purchasing power of American's middle and blue collar classes and increased everyone's tax burden for public assistance benefits.

Publicly owned enterprises, whether in a socialist or a capitalist country, have similar problems, even when showing a profit is not required. They have fixed budgets, must contain costs to avoid public outcry, may have quotas imposed on their production, must rate employee performance, and must also downsize when directed to do so, as many agencies recently did. Public enterprises may also have an obligation to provide goods and services to people who cannot afford to pay for them.

Other Benefits of Development

While the majority of people today seem to support economic growth in principle, many of them favor particular types of development and oppose others, e.g., a new oil field, a large industrial plant, a regional mall, a major freeway extension, a new prison, or an order from the Pentagon for 50 more planes. Proponents of each type of development usually extol its potential benefits and downplay its costs, whereas opponents tend to emphasize costs.

When businesses expand and bring sizable new payrolls to a community, additional local and regional economic growth is expected to follow. When jobs are created, some family incomes increase and public outlays to the unemployed are reduced. Some consumers will be able to afford their first home, a better auto, an appliance, new clothes, more groceries, a dinner out, or toys for their children. In time, state agencies, counties, communities, and school districts collect increased revenues from taxes on income and property, and more fees are collected.

But the economic stimulus doesn't end here. The companies doing the work generally need more materials and equipment, make some local purchases, offer subcontracts for work, and may increase payments to their stockholders. Hence, some local and regional businesses make additional sales and those that expand their facilities or offer new services often hire more employees. This further increases local spending, encourages the establishment of a few new retail sales or service firms, and again increases public revenues. Local residents are then more apt to view their community as progressive, to boast of their population increase, and to welcome the additional consumer choices that growth has provided.

Potential Social and Economic Costs

Opponents of development are likely to see the potential effects of another oil field, an industrial plant, a regional mall, or more warplanes from a different perspective, especially if it is rapid and unplanned. Jobs are created, but often the best positions are filled by corporate transfers and skilled in-migrants, leaving mainly semiskilled, low-wage openings for local workers. When higher pay is offered, competent local employees are hired away from their jobs and older community businesses and services suffer. Local facilities and services are overburdened and older residents find this inconvenient if not distressing. Often new equipment and supplies are purchased from distant sources and subcontracts go to firms in other cities. Growth also attracts new chain stores and restaurants that displace local businesses.

When the economic boost is a temporary project (e.g., an oil field or a freeway section), the jobs are short-term and most workers may be single transient males who do not always respect community customs and traditions. New construction projects sometimes force relocation of residents, inconvenience others with noise, traffic disruptions, or visual blight, and consume energy and other nonrenewable resources. Project activities sometimes degrade the environment and this may cost the taxpayer more than the developer in the long run (Ch. 5).

Long-term economic expansion (e.g., the new industrial plant or regional shopping mall) attracts in-migrants who require more schools or classrooms, hospital space, police and fire protection, streets and parking space, and water and sewer facilities. More tax revenues are needed to support this expansion. The demand for housing increases but so does the cost of housing and perhaps other essentials for everyone in town, including people on fixed incomes. If the economic venture is a risky one and fails within a few years, the community is stuck with surplus facilities, increased indebtedness, and no revenue increases to pay for them.

Two types of agreements between local governments and developers are becoming more common. If a county or community is eager to attract a new industry, it may offer incentives, such as a change in zoning restrictions or a reduction in property taxes. Sometimes cities compete to see who can lure a major industrial plant with the best package of benefits. Alternatively, if a corporation or agency is anxious to overcome local resistance to a project, it may offer to build a new school or library, to provide low-cost housing for its workers, or to hire a high percentage of local residents.

Thus industrial development has both its environmental and social price tags, leading a growing number of people to wonder about the feasibility of perpetual economic growth and whether we have already gone too far or too fast (Ch. 4-5).

Economic Growth and Social Change

Economic growth transforms entire societies as well as individual communities, especially in developing countries (Ch. 1). Major changes occur in the type, scope, and distribution of economic activities; in the structure of social institutions, such as families, churches, and the government; and in the beliefs, values, and norms of the people.

Technologically Induced Change

Successive "waves" of change seem to occur following important technological innovations, as they do following other major events, such a world war, a deadly pandemic disease, or a comprehensive free trade agreement. The factory system reduced self-employment and removed most economic activity (also fathers and now mothers) from the home. Then agriculture was mechanized and tractors, threshers, gang plows, mechanical feeders, milking machines, harvesters, and other machines greatly diminished the need for farm labor.

In the United States, where productive efficiency in agriculture has been increasing for generations, farms now employ about 2.5 percent of the population, in contrast to 53 percent in 1870. In the interim, many small farms have failed and much cropland has been merged into larger holdings. Numerous farmers and their children have been forced into other employment despite two generations of federal programs to sustain the family farm. Similar trends are now evident in other nations worldwide.

The advent of the automobile displaced dozens of previously important businesses, such as livery stables, carriage and harness makers, and dray services. It created new jobs in auto plants, garages, gas stations, road and bridge construction, drivers' training, and an array of drive-in services. Together with paved roads, the popularity of the auto also led to the demise of most small towns. People preferred to shop in larger towns where the selection was greater and prices were often lower. The speed and ease of riding decreased the number of pedestrians on the streets and highways, increased opportunities for urban crime, and reduced walking for exercise.

We are now in the midst of an electronics revolution that is dramatically affecting both work and leisure roles in homes, schools, offices, and factories worldwide. The personal computer, electronic mail, and now the information highway are but the latest phases of a larger transformation that includes radio, television, stereo sound, compact discs, and factory automation (Ch. 1). Traditional jobs are being eliminated, while new opportunities (often fewer in number and requiring more training) are created. Supervisors compose and send their own letters, lovers communicate via E-mail, robots work on assembly lines, children play computer games rather than physically active activities, and parents are concerned about the potential of the new media for vulgarity, pornography, and bad habits.

Urbanization and Its Impact

The concentration of industry near transportation hubs, cheap power sources, supplies of raw materials, and/or markets has drawn people to the cities. In 1992, 80 percent of Americans lived in 269 metropolitan areas and the majority of the people in most other nations also live in urban settings. Widespread urban values and norms, such as employment away from home, maximizing wealth, enhancing social status, following fads and fashions, increasing family comfort and convenience, relying on legal contracts, weakened extended family ties, minimal neighborhood interaction, and relentless consumption, are steadily replacing traditional rural perspectives.

Rural Americans, more often than urbanites, were proud of their economic independence, had few social distinctions among community members, operated more self-sufficient households, had strong family loyalty, got by with what they had, shared resources with others in time of need, and honored handshake agreements.

Strong national governments, compulsory education, efficient transportation and communication, mass marketing, and rural to urban migration have forged nations from earlier mosaics of ethnic groups separated by barriers of topography, language, religion, and social status. The same influences, together with freer trade and the quest for profits, are now lowering the barriers among nations and making them more dependent on each other. Western news media, magazines, fashions, films, television programs, eating habits, and other consumption patterns influence people all over the world, while foreign products and tastes also make a significant impact on the industrial West. International trade has increased at least twentyfold since 1950 and nonwestern factory goods are a growing share of the total world market.

Cultural integration is now relentless and occurring on a global scale, as evident in the use of English as the major language of commerce and tourism, adoption of the metric system in most of the world, international programs to improve health and literacy, and increasing standardization of products and parts. These trends disturb many people who are proud of their heritage and fear the loss or dilution of ethnic unity, religious doctrines, and cherished traditions (Ch. 10).

Many people welcome these trends, believing that greater international understanding and cooperation are essential to solving the problems that now confront the world. Note the contrast between the relative harmony and prosperity of the European Community today and Europe's destructive past. Compare this with the conflicts and carnage of Bosnia, the Middle East, and central Africa where traditional animosities prevail.

"Green" Enterprises

The expanding knowledge and sophisticated technology that stimulated so many social changes and environmental problems also have the potential to mitigate many of these conditions. We use new technology to monitor population characteristics and to diminish the rate of growth. We also use it to determine natural resources needs, to estimate remaining reserves, and to assess and reduce environmental degradation (Ch. 8).

The economic sector (private enterprise) has discovered that protecting the environment can be profitable and has responded in several ways. Various private firms now offer a wide range of equipment and services, including scrubbers that remove toxic chemicals from plant emissions, bacteria that fight plant pests, recycled paper, renewable energy sources, systems for recyling used materials, equipment and techniques for decontaminating plots of land and water bodies, and consulting services to assist agencies, corporations, and communities.

Engineering consulting firms assess existing and potential environmental degradation from development activities and recommended needed action. Others prepare required environmental documents or take remedial action to reduce environmental impacts. A growing eco-travel industry features tours to environmentally critical areas to observe conditions first hand. Some companies that have historically wasted resources and created pollution are now touting their efforts to replant logged tracts, to recycle materials, and to reduce environmental degradation.

In Germany, a world leader in this technology, bins are available throughout each city for depositing discarded paper, glass, metal, and organic materials. Industries extract pollutants from their plant emissions and use these to manufacture useful products, such as wallboard and other construction materials. Stern environmental regulations plus the high cost of fossil fuels have encouraged innovations, including recycling solvents, reducing waste, and using alternative forms of energy (1). Resource-poor Japan is also minimizing waste, especially of nonrenewable resources.

Similar efforts that were initiated in United States during the 1970s under President Nixon suffered a setback when the Reagan administration de-emphasized the need for conservation and environmental reform, and sharply reduced financial support for these federal programs (Ch. 8, 10).

These examples are just the tip of the iceberg as nations get more serious about environmental needs. Many more jobs could be created in environmental education, improving and extending public transit, increasing energy efficiency, recycling additional materials, and restoring forests, streams, lakes, and wetlands.

Persistent Social Issues

The Legacy of Colonialism

Until the mid-1900s, military conquest and colonization were common practices in much of the world. Powerful nations with strong central governments, superior technology, expansionist ideals, and capable armed forces dominated other lands. Foreign countries and territories were coveted for their agricultural lands, strategic location, valuable minerals, abundant timber, animal pelts, cheap labor, and potential as a market.

At various times during the past three centuries, a dozen nations (Great Britain, Spain, France, Portugal, the Netherlands, Belgium, Germany, Denmark, Italy, Russia, Japan, and the United States) dominated the governments and economies of much of the rest of the world. Colonial areas were controlled by several means, including military force, subversion, aggressive trade policies, installing puppet governments, large-scale emigration to them, suppressing local customs, annexation, and religious conversion. Colonial powers sometimes justified their imperialism with claims that subjugated peoples were culturally backward or intellectually inferior, their resources were going to waste, or they could benefit from western influence.

Just a century ago, the British Empire alone included about one-fourth of the world's land area and population, by far the largest empire the world has known, and British citizens were just two percent of the population. Portugal, France, Spain, and the Netherlands also had empires much larger and more populous than their own country. During World War II, Germany (the Third Reich) briefly controlled much of western Europe and Japan dominated Korea, parts of China, and large areas of the North Pacific. Until its recent break up, the Soviet Union governed many disparate ethnic groups, dominated most of eastern Europe, and strongly influenced the governments of several other countries around the world.

Foreign powers often exploited indigenous peoples and their resources, belittled their customs and religion, and generated resentment that contributed to their eventual ouster. World opinion now discourages colonialism and relatively few people are dominated against their wishes by clearly foreign powers. Yet serious rivalries and civil strife still occur among indigenous ethnic groups and factions, as in Bosnia, Ethiopia, Rwanda, Iraq, India, Northern Ireland, Korea, and Viet Nam.

Even though most former colonies are now independent, the impact of colonialism is still evident. In these societies, many key positions were formerly held by foreigners, old social structures were weakened, education was neglected, and citizens were unprepared for self-rule. At the same time, colonialism was a vehicle for transmitting many traits of western culture worldwide and its institutions and modern infrastructure often serve as models for social and economic development in newly independent nations.

Modern communications media and a variety of international activities continue to influence these nations. The United Nations, various national governments, and private groups offer grants, loans, material aid, health services, and technical assistance. Multinational corporations develop mineral and fossil fuel resources, cut timber, and build factories. Corporations often hire foreign workers for a fraction of the wages paid in their home country and may be subject to fewer or poorly enforced environmental protection procedures.

Defenders of overseas operations say they are reducing prior unemployment and poverty and expanding foreign markets for U.S. goods. Their critics cite examples of grim working conditions, child labor, and toxic pollution emissions that may jeopardize the health and welfare of employees and local residents. Host countries permit these actions to create jobs in their burgeoning cities, to expand their tax base, to obtain western currencies to implement their own projects, or to secure other benefits for public officials and business interests in their country.

Status of Women

In most countries of the world, women have a more restricted legal status than men, and in all of them there are sex-related differences in roles and career options. Even in the most progressive societies, some obstacles remain to deny women certain opportunities and privileges available to men. Many people now believe that sexual inequality is dysfunctional to the larger society as well as frustrating to women who encounter these barriers.

When women are subservient to men and are denied equal educational, career, and leadership opportunities, their country loses some of its best potential for research, problem solving, artistic expression, and project management. Well-educated women with varied career options usually make valuable contributions in businesses and the professions, play leadership roles in their communities, limit children to the number they want and can afford, and take steps to ensure that their families are healthy. They are also able to leave abusive relationships through divorce or separation and to support themselves.

Ethnocentrism

Many of the social problems facing humanity today are global in scope, including nuclear weapons control, international terrorism, accelerated population growth, diminishing natural resources, and environmental degradation. Yet the world is still divided by religious, ethnic, and ideological rivalries that often impede international negotiations for solutions to these shared problems.

Sociologists use the term ethnocentrism to denote the tendency of most people to regard their particular ethnic traditions, cultural values, religious beliefs, economic system, and even foods, dress, and other aspects of social life as more normal, reasonable, and defensible than those of other cultural groups. Ethnocentrism is both functional and dysfunctional to social life. It is the "social glue" that binds an ethnic group or nation together and encourages unity, conformity, loyalty, and pride. It is also a major basis for animosity, ridicule, discrimination, and aggression against other groups.

Extreme feelings of superiority or righteousness make it easier to justify cruel acts toward others who are racially or culturally different by accident of birth, even when they are good citizens. Ethnocentrism often discourages close cooperation among groups with common problems, such as the need to curb population growth, to reduce civil strife, to exchange technology, or to share resources.

In the United States and many other nations, population policy and natural resource depletion are just two of the areas of strong disagreement that are rooted in religion and ideology. Many conservatives want to reduce the government's role in conserving and preserving natural resources, protecting the environment, and supporting family planning, especially the use of abortion. Most liberals think we need these and perhaps more safeguards to curb excessive population growth and to ensure adequate resources and a healthy environment in the years ahead (Ch. 8-10).

Antisocial Behavior

Antisocial behavior includes a wide range of hostile acts that cause pain, grief, or anxiety, or wrongfully deprive victims of their possessions. Motives include greed, rage, revenge, ignorance, and sadism, as depicted in our daily newspapers. Common examples of antisocial behavior in the United States and elsewhere are murder, assault, robbery, burglary, arson, tax evasion, spouse and child abuse, and illegal drug sales. There are also so-called "crimes without victims" (that usually do have victims), such as drinking to excess, "recreational" drug use, illegal gambling, suicide, and prostitution.

Many of these acts occur in all societies and eras, but vary considerably in their frequency and intensity. Environmental crimes (ecocrimes) are apt to become more common as the ratio of people to resources increases and more regulations must be imposed to ensure environmental quality. Examples include timber theft, unauthorized use of water, trespassing, illegal dumping of pollutants and toxic wastes, and poaching wildlife.

In the United States, several types of antisocial behavior have been increasing for decades, as if they were by-products of industrial development, urbanization, and expanding consumption. Official statistics showed per-capita increases in murder, rape, assault, robbery, drug abuse, larceny, and auto theft, with more modest increases in the last decade or two (2). Part of the increase was attributed to improved reporting and recording procedures, to the increase in the percentage of teenagers and young adults following the "baby boom" of the 1950s, and to increased drug addition.

U.S. prison populations have increased dramatically, especially due to drug-related crimes and longer sentences (3).

Official statistics tend to focus on the so-called conventional crimes mentioned above, but white collar and corporate crimes are also widespread. These include embezzlement, stock and bond swindles, insider trading, industrial espionage, price fixing, false advertising, unauthorized mining and logging, and violations of environmental restrictions. Then too, we have criminal syndicates that are now highly organized, international in operations, and professional in their methods.

Although explanations for rising crime rates vary, some impressions are widely shared. To begin with, the constant barrage of multimedia advertising is thought to raise our levels of need and expectations. Wealth and possessions are important measures of social status and accomplishment. Modern societies are also very "open" in the sense that informal controls on people's behavior are relatively weak, especially when they are among strangers. People in the United States are free to move anywhere and are neither required to carry passes nor to inform local authorities when they move to a new location. Traditional sources of social control, the family, the church, the school, and the employer, are less effective than formerly, as rival influences gain strength and individual rights are increasingly respected and defended. Some recent symptoms of family and community instability are rising rates of divorce, teenage pregnancies, alienation of the elderly, youthful runaways, and homelessness.

Young people today have more uncommitted time, income, freedom of choice, and access to cars than a generation or two ago. They also have less supervision, spend more time away from home, and find it easier to ignore laws and social conventions. Broken homes are far more common today and many children are reared with little supervision from fathers. Personal autos give young people autonomy and instant anonymity. The influence of peers remains strong, however, and many people, both old and young, are willing to observe "situational" ethics when obeying conventional ethics would displease their friends or coworkers. The growing abundance of goods and services is a major factor in explaining the increase in property crimes. When your grandparents were teenagers, they couldn't possibly steal a computer or television set, pirate a recording, or put a virus into a computer program, since these opportunities did not yet exist. Neither could they easily obtain marijuana, sniff cocaine, or carry a noisy portable radio down the street, and they may have lacked the time and money and to commit many other misdeeds.

Illegal Immigration

The lures of greater wealth and personal freedom continue to encourage many people to emigrate to the United States and other prosperous countries. People officially classed as refugees have increased from 2.5 million in the mid-1970s to 23 million in 1994 and additional millions are migrating for other reasons (4). If they had the means and opportunity, far more people would like to escape extreme population density, dictatorial governments, miserable living conditions, deteriorating local environments, natural disasters, and limited jobs and income in their native countries.

Following the Great Migration of 1820-1930, primarily from Europe to the United States and Canada (Ch. 2), a "second wave" of immigration beginning about 1940. A growing majority of the 19 million legal immigrants to the U.S. are from Mexico, the Caribbean, Central and South American, and south and east Asia (5). Subsequent births and concurrent illegal immigration have added more millions, and the total national population has doubled from 132 million in 1940 to 265 million in 1996. Many nations have established annual immigration quotas (Ch. 2) in response to mounting public concerns about overcrowding, unemployment (especially among the unskilled), education, and the cost of social services. U.S. quotas permit close relatives to join their families and also admit some employable people who seek permanent residence and citizenship. Usually legal immigrants must meet certain basic requirements for admission as candidates for citizenship. People are also admitted when they can prove the need for asylum from political persecution. According to an October 1996 CBS news report, 1.1 million immigrants became U.S. citizens within the past year.

Millions of people enter various countries illegally, some by obtaining temporary permits and staying on, and others by sneaking in or being smuggled. Many of these immigrants are potentially hard working laborers seeking temporary employment to acquire a nest egg to take back home. Others hope to remain undiscovered and ultimately qualify for amnesty and permanent residence or citizenship. Some are criminals and opportunists who use the increased personal freedom and opportunities in a democratic country to conduct illegal activities.

Critics of our past tolerance of illegal immigrants in the U.S. claim that four million are already here and that another 300,000 or 400,000 arrive each year. They say that illegals cost us more than they contribute and should be denied most social services. As with legal immigrants, the largest number come from Mexico, followed by Central America, the West Indies, and southeast Asia. If population pressures and poverty increase in these areas, one might expect even more illegal immigration.

A lively debate has ensued about whether immigrants cost the nation more than they contribute, with each side citing evidence to support its claim. Population-Environmental Balance, a Washington-based public interest group, expresses concern about present U.S. immigration policy that legally admits about a million people annually, five times the number that emigrate from the U.S. to other countries. Balance believes and many others agree that legal and illegal immigrants depress wages below an adequate income for America's unskilled workers, cost billions more in social services than they pay in taxes, and increase already excessive population density and environmental degradation in some areas.

Other sources including the conservative Cato Institute dispute this, saying that immigrants contribute more to our society than they collect in social services. Part of the discrepancy seems to be the different measures used to determine both their contribution and their burden to taxpayers, as each side strives to make its case. A 1994 New York Times-Mirror Center poll indicated that 82 percent of Americans support immigration restriction, up from 76 percent in 1992. A recent CBS-Times poll showed that over 60 percent of Democrats, Republicans, and independent voters favor reductions in legal immigration.

Role of Poverty in Population Growth

Extreme poverty plays a mixed role in the world population explosion. Among the poor, death rates, especially for infants and the elderly, are higher than in the general population due to malnutrition and substandard health care. The poor also tend to have much higher birth rates than the middle classes, frequently doubling their numbers in a single generation. Compared to other social categories, the poor are more often ill-informed about effective birth control measures, lack the cash to buy pills and contraceptive devices, and ineffectively use donated birth control devices. Many also feel obligated to observe religious restrictions on birth control.

In the United States live births among unwed mothers tripled between 1970 and 1992, even though total births to all mothers increased only eight percent during this period. The highest birth rate (95 per 1000 women) was among women with a family income of less than $10,000, followed by mothers (70 per 1000) with a family income under $20,000. Mothers in the highest income bracket, $75,000 or more, had the lowest birth rate (43 per 1000). Births to unwed teenaged mothers almost doubled during but most of the fathers were twenty or older (6).

In many countries, including the United States, the poor are disproportionately racial or cultural minorities coexisting with "mainstream" groups that dominate the political, economic, and religious institutions. The poor may also, as in the Amazon Basin and parts of Africa, be members of indigenous tribes that are isolated from or weakly integrated into larger societies and eke out a living from traditional activities, such as gathering forest resources, fishing, herding, and subsistence agriculture. In the colonial era and in sub-Saharan Africa today, most citizens of entire nations are poor by global standards. Their institutions are controlled or influenced by foreign interests, often in cooperation with self-serving local nationals, and much of their national wealth is diverted to relatively few people.

Environmental Justice

Economically disadvantaged people tend to be relatively powerless politically and less-informed than other citizens about impending events that will affect them. When they offer little organized resistance to changes they are easy victims of industrial development; e.g., freeways that bisect their neighborhoods, plant sites that emit pollution, removal of forests, mineral development, high dams that flood their homelands, exposure to pesticides, or landfills that produce odors and contaminate wells.

An environmental justice movement is gaining strength in many nations as potentially impacted residents organize to protect their interests. Human rights groups and environmental activists, sometimes at great personal risk, also support the cause of disadvantaged groups. The underlying assumption is that the needs of all people who would be affected by proposed industrial developments should be considered before a decision is made and that the outcome should reflect this process. Sachs describes examples of community, regional, and international efforts to ensure environmental justice, and concludes that these measures ultimately protect environmental quality for everyone (7).

Business Cycles

Uneven economic development in industrial countries produces business cycles with alternating periods of industrial expansion and retrenchment. Each phase often lasts one to three years but is moderated by government regulation of financial transactions including interest rates and investments, and by the business sector itself, which adapts to changing market conditions.

A spectacular example of a business cycle is the 1920s economic boom in North America and Europe and the Great Depression that followed in the 1930s. Factors that contributed to the Great Depression in the United States include the mechanization of agriculture, expanding factory outputs, easy credit policies, an extended period of drought for farmers, rapid population growth, and the very unequal distribution of wealth. The great majority of the population had limited purchasing power, farmers could not sell their products at a profit, and automobile plant employees were unable to afford the vehicles they produced.

The result was overproduction of goods, a devastating stock market crash, and an extended economic depression in which the U.S. gross national product fell from $104 billion in 1929 to $56 billion in 1933. North America and Europe experienced massive product surpluses, industrial phasedowns, plant closures, mortgage foreclosures, and widespread layoffs. Fortunes were lost, U.S. unemployment exceeded 30 percent, and many farmers lost their lands and machinery to creditors. Food lines became a daily sight, farmers sometimes had to sacrifice their livestock and equipment, and barter their products for foods and services. Average family size declined sharply, used clothing passed from one family to another, children stayed in school longer, and thousands of jobless hoboes roamed the countryside.

Ultimately the cycle was broken by various federal job programs (e.g., WPA, PWA, and CCC), by state and private efforts to increase employment and to distribute aid to the needy, and by defense production and mobilization on the eve of World War II. Since that time the federal government has more carefully regulated the economy to minimize unwise investment practices, inflation, and long-term recessions.

Today there is widespread support for economic growth to provide jobs and other necessities for expanding populations and to improve the living standards of developing nations. But we see growing sentiment the planet cannot support limitless population and industrial growth (Ch. 8-9). Left unanswered is whether economic growth is really necessary to maintain a reasonable level of employment and prosperity when zero population growth has been achieved, as in much of Europe. The jury is still out on this issue and the answer may depend largely on the effectiveness of economic and fiscal policies of individual nations. Social and environmental awareness, intelligent analysis of resources and needs, careful planning for the future, and public and corporate responsiveness may lead to a satisfactory solution.

Chapter 7

Agriculture, Nutrition, and Health Issues

One of the tragedies of the modern era is that we now have the means to eliminate or reduce most of the hunger, deprivation, and disease in the world but have not done it. As of 1995, the world produced enough food and materials to feed, house, and clothe each family. There are drugs to combat most infectious diseases and we could eliminate some of them. We can avoid having unwanted children without resorting to abortion and could have a sustainable world population if we could agree it is essential. Finally, there is enough important work to employ the jobless and enhance their self-respect, stimulate economies, increase tax revenues, diminish the need for social services, and reduce crime and drug abuse.

All nations fall short of such utopian goals, although some societies clearly outperform others in particular areas. But no society meets all human needs well, as suggested in Table 7-1.

Table 7-1: Barriers to Social and Economic Well-Being

Obstacle	Examples
Social conditions	Concentration of poverty in some depressed areas, high illiteracy, lack of marketable skills, ineffective family planning
Cultural fences	Ethnic rivalries, rigid class structure, religious or ideological opposition to needed changes
Economic practices	Priority on profits, minimum labor, short-term gains, very unequal distribution of earnings from production
Resource imbalances	Limited forests, minerals, croplands, or water in relation to population, harsh climate, lack of ocean ports, limited access to markets
Technological lag	Insufficient or obsolete industries, inefficient farming methods, meager research and development, inadequate transportation, communication, utilities, schools, health services
Antisocial values	Greed, nepotism, discrimination, hedonism, opportunism, irresponsible role models
Political policies	Priority to favored elites, restricted personal freedom, widespread corruption, use of terror
Personal attributes	Unaware of personal potential, mental or physical impairment, grim resignation to conditions, lack of self-discipline

Global Food and Health Needs

Estimates vary, but about one-half billion people, over a tenth of the world total, are starving or suffering from severe malnutrition and as many more lack enough food to maintain their vitality (1). The greatest concentration of hunger is in Africa, followed by Asia, and then Latin America. Even though official estimates may omit other regions, some people in every country lack sufficient food. Many nations consume all they produce, have few or no reserves, and face the prospect of famine if diseases and drought reduce their annual harvests.

Untold millions of people, including some Americans, are sick or dying from lack of medical care. As more people are displaced from subsistence agriculture by population growth, depleted croplands, and competing land uses, they must have an adequate cash income to survive. Yet millions of jobless around the world valiantly seek but are unable to find employment or earn a wage too low to cover the most basic expenses.

Clearly none of these conditions is new and critics of population policies and the environmental movement often point out that humanity never had it so good (Ch. 9). They note that we live longer, produce more food, have better medical care, enjoy more consumer options, and have improved air quality over many cities. In prosperous countries, there are welfare and pension benefits, economic and educational development strategies, and the potential of modern technology, including biotechnology, to resolve new problems as they emerge. Aid to needy nations is now available from national, international, and private food and medical aid programs.

The validity of these claims does not alter the fact that life is hard for a sizable portion of the world's people and prospects for many look very grim. Recent developments in agriculture and health provide clues to what lies ahead and some innovations that could modify present trends.

Agricultural Trends

Around 1970, several experts evaluated population trends and predicted that massive famines would occur in many developing nations within a decade or two. But in the mid-1980s, starvation was much less extensive than expected, especially in southeast Asia, and in some countries the situation had improved. The Green Revolution and various foreign aid programs kept millions of people alive that might otherwise have perished and in most countries populations continued to increase (Ch. 2).

The Green Revolution

The Green Revolution refers to the adoption of prolific new strains of grains and other food crops that require large quantities of fertilizer, pesticides, and water but produce remarkably high yields. This revolution was initiated in the 1960s by a team of agricultural scientists headed by Norman E. Borlaug, who developed high-yield wheat, maize (corn), and later rice. These grains were quickly adopted by developing nations desperate to feed their people and between 1965 and 1980, world grain production rose 55 percent to 1.57 billion metric tons.

Other food production also increased, even though at lower rates than grains, but these trends soon leveled off. Using World Resources data, Earth's population grew about 60 percent between 1965 and 1990, while grain production increased 80 percent. Meanwhile fruits and vegetables rose 73 percent; meat, milk, and fish were up 60 percent; and root crops gained 20 percent.

In the past decade, these rates of growth in agricultural production have decreased. Between 1980 and 1993, according to the 1995 *Statesmans Yearbook*, world grain production increased only 20 percent to 1.88 billion tons. The largest percentage gains were in rice (33) and wheat (29), while production of maize (corn), rye, barley, oats, maize (corn), rice, millet, and sorghum were lower (11). World population increased 25 percent during the same period. Current Worldwatch data indicate a decline in world grain reserves for 1996, the lowest in decades, due in part to the hottest summer on record in 1995. As food production levels off and the number of people continues to increase, there is less food per capita. Worldwatch Institute's books, *Full House* (1994) and *Who Will Feed China?* (1995), provide further insights into this decline, based on United Nations FAO and USDA data up to 1993. Per-capita grain production decreased 12 percent since 1984, seafood harvests declined 9 percent since 1988, and beef and mutton dropped 13 percent since 1972 (2). The Chinese government has since announced some improvements in food production.

There are several reasons for the world decline in per-capita food production. Green Revolution technology is now widespread (thus stabilizing production), fresh water use is approaching the limit, and benefits from increased application of chemicals are diminishing. Available croplands and fisheries are now quite fully exploited and reduced in some areas by pollution, land erosion, and the competing land uses of growing populations. Although yields of some plant species continue to be improved, no major revolution in agricultural technology is anticipated in the near future. Yet population growth in most developing countries continues to exceed the world average.

Environmental Effects

It is certain that increased crop yields of the Green Revolution have averted or reduced severe famines in many countries. Critics regard the revolution as a temporary, environmentally risky strategy and propose additional measures, such as improved technology for really small landowners. They note that water supplies are limited in most of the world and that many poor farmers lack the capital, skills, and acreage to profit from the new grains. In addition, some of the liberally applied fertilizers and pesticides are washed from the fields by heavy rains and contaminate streams, groundwater, and ultimately lakes, estuaries, and bays.

The nutrients and the sediments that wash into streams and lakes create lush vegetation that may eventually stifle the flow of water and raise its temperature, creating boggy areas, killing fish, and eliminating smaller lakes (Ch. 5). As the new strains of grain replace older varieties, genetic diversity is drastically reduced, making the annual crop of wheat or rice more susceptible to failure from insects and parasites. In a grain-dependent country, the loss of just one season's crop of these new varieties could be devastating.

These problems are not limited to the developing world. Barry Commoner has calculated that between 1950 and 1987, while U.S. crop and livestock production increased 80 percent and farm labor decreased 71 percent, the use of insecticides, herbicides, and fungicides increased 484 percent (3). One can readily imagine similar trends in other industrial countries.

The Persistence of Poverty

Compared to even a generation ago, many more people now survive childhood and become parents due to the Green Revolution, improvements in food preservation, expanded marketing and distribution services, and public and private food and medical aid programs (Ch. 2). But in the persistently poor nations of Africa, Asia, and Latin America, where population growth exceeds economic growth, they often exist at a bare subsistence level. Most of the 35 countries from the Sahara Desert into south Africa have high birth rates, declining per-capita food production, diminishing soil fertility, and extensive malnutrition. Their problems are compounded by their inability to extend croplands, by high levels of unemployment, by the prevalence of fatal or disabling diseases, and by ethnic rivalries and civil strife, as in Ethiopia, Sudan, Somalia, Angola, Nigeria, Liberia, Rwanda, Burundi, and Zaire.

Limitations on Croplands

Contrary to what a lot of people think when they see large expanses of "vacant" grasslands, forests, and deserts, most of the land on earth that is suitable for agricultural production and many marginal areas are already under cultivation. Cropland and pasture expansion is limited by factors such as climate (rainfall, growing season), terrain, soil type and fertility, available fresh water for irrigation, accessible markets, and the need to conserve remaining forests and wetlands.

In Latin America, 10 percent of the population owns most of the land and many farmers do not own the plots they till. Hundreds of millions of Asians and Africans farm on a very small scale by U.S. standards and under difficult conditions. They lack the capital to make improvements, are subject to crop failure from floods and droughts, and lack the acreage and water to utilize high yield grains or modern farm implements efficiently. Many cannot produce enough surplus to sell or barter to meet their other needs.

Worldwatch Institute observers believe that we are entering a new era in which industrialization and other needs in developing nations will claim more cropland, while food surpluses for export will be more limited. According to the *Statesmans Yearbook*, three percent fewer acres of grains were planted in 1993 than in 1980, reflecting a loss of croplands and the difficulty of adding new plots. As densely populated countries, such as Japan, China, Taiwan, and South Korea, industrialize, their loss of croplands to other uses reduces food production and they must import what they lack. Since 1993, China, with 20 percent of the world's people and only about seven percent of its croplands, has been a net importer of grains. China purchased 4.3 million metric tons, mostly corn, from abroad in 1995, chiefly from the U.S. Midwest.

In developing countries, the many competing land uses include human settlements, industrial complexes, water reservoirs, corridors for roads and utilities, mines and tailings ponds, timber plantations, parks, protected wetlands, game refuges, and forest preserves. The construction of high dams to generate electrical power for new cities and industries submerges valuable croplands and displaces thousands of rural residents.

Importing food is of course very feasible when exporting nations like Great Britain, Belgium, Japan, Singapore, and Saudi Arabia have the money to buy it and other countries have surplus food to sell. This is presently the case, but if more nations need to import foods and fewer of them have surpluses to sell, prices on the world market will rise and encourage cultivation of more marginal lands. Some lands are much more suitable for forests or grazing than for grains or garden produce, but rising food prices would encourage more conversion of woodlands and pastures to croplands despite the expectation of low yields.

Livestock versus Grains

A loss in productive potential occurs when fertile croplands are used to raise livestock because more food could be produced if the plot were used for cereal grains and vegetables which are lower on the food chain (Ch. 5). Compared to livestock, many food crops require less commercial energy and water, produce fewer solid wastes, and do not compact the soil and degrade stream banks the way animals do. But the world demand for meat in the past 50 years has increased twice as fast as population growth, due in part to increased prosperity in many industrial countries. The U.S. leads with annual per-capita consumption of about 240 pounds of meat, some 50 times more than India at the other extreme. Accordingly, about 70 percent of U.S. grain is used to feed livestock, compared to only 2 percent in India (4).

Livestock require not only space for pastures and feedlots, but additional croplands to produce their feed. China has an estimated 350 million pigs, over five times more than the U.S., and two billion chickens. India, with a third of the U.S. land area, three times as many people, and very limited pasturage, has twice as many cattle (almost 200 million), over 100 million goats, and 74 million water buffalos. In many countries animals have important uses in addition to supplying meat, eggs, and dairy products. They provide transportation and pull farm equipment, are a source of fiber, pelts, fuel, fertilizer, and building materials, and meet the sacred or emotional needs of people.

The Paradox of Population Technology

Technology has increased the average life expectancy of newborn humans worldwide (Ch. 2). In the industrial nations, expected longevity was extended from about 50-55 in 1920 to 75-80 today. In many developing nations, increases during this period were even more dramatic, from as few as 10 to 20 years to 40 to 70 years now. Most children today live long enough to become parents and grandparents, and frequently three different generations must share their property and living space. In the most populous regions of the developing world, many family farms are small (often less than an acre) and can no longer be subdivided to give another generation sufficient land to support children adequately and to get them established in other careers. Declining soil fertility from repeated plantings of the same crops, wind and water erosion, compacted soils, and pollutants is another limitation. In desperation, some landless farmers burn portions of forest reserves to clear land for farms and cattle ranches. A 1995 Associated Press report stated that smoke from lowland fires in Bolivia covered two-thirds of the country and posed a serious threat to both the environment and the health of residents.

Other farmers plant crops on unsuitable lands, such as deserts, floodplains, or steep terrain, in the hope that droughts, floods, or landslides would not prevent a harvest. Too often these ventures fail and the lands, stripped of cover, become severely eroded, reducing their ability to revert to native vegetation and wildlife.

Many developing countries experience an initial population boom, doubling their numbers each generation as health standards improve and food supplies become more dependable. The same effect once occurred in each "older" industrial nation and the growth cycle continued until a large, well-educated middle class emerged, perceived advantages in small family size, learned how to limit births, and then rejected the large-family tradition of previous generations (Ch. 2).

In many developing countries today, poverty and illiteracy are widespread and this middle class is slow to develop. Birth control strategies are often ineffective and population growth continues at an unsustainable rate, straining local resources. The USDA estimates that the demand for emergency food aid, not now fully adequate to the need, will double within another decade, most assuredly in sub-Saharan Africa.

Displaced young people often move to the cities to seek employment and find that the labor supply exceeds the demand and that wages are depressed when work is available. Many nations seeking to establish new industries to support the new generation are handicapped by lack of capital, limited literacy and technical skills, restrictive traditions, ethnic rivalries and warfare, and political corruption.

Modern Health Hazards

Despite the medical advances of the past 125 years and the sharp upward trend in life expectancy in virtually every country, there is mounting global concern about insidious malnutrition, a resurgence of disease epidemics, increased risks from toxic chemicals, and possible health effects from global warming and the thinning ozone layer.

Malnutrition

The U.S. Department of Agriculture estimates that at least a third of Americans have inadequate diets. Sometimes this is due to poverty, but more often it reflects a poor choice of foods, with too much emphasis on soft drinks, fried foods, rich desserts, and too few fruits and vegetables. U.S. citizens average 47 gallons of soft drinks annually compared to 24 gallons of milk and 8 gallons of fruit juices. They eat 65 pounds of fats and 147 pounds of sugar and other sweeteners each year. About 28 percent are now 20 percent or more overweight, even though just 56 percent eat breakfasts regularly (5).

Thus malnutrition is a relative term and perhaps half of the world's people would be so classified if we used the USDA's lofty guidelines for an appropriate diet. Applying a more frugal standard, about one billion of the globe's population suffer from severe malnutrition because they lack the quantity and variety of food needed to ensure normal growth and body weight, adequate role performance, and a healthy immune system. Aside from the stress and discomfort of food shortages and uncertain meals, these people sometimes lack the strength to work, are more susceptible to diseases and injuries, and may resort to begging or stealing to survive.

Threat of Epidemic Diseases

A generation ago, health authorities believed that most dread diseases could in time be eliminated. Their weapons included vaccines to confer immunity, antibiotics to kill bacteria, pesticides to control virus-carrying insects, and improved access to previously remote areas of the world. During the 1950s and 1960s, impressive gains were made in suppressing ancient epidemic diseases, such as smallpox, polio, cholera, bubonic plague, tuberculosis, and yellow fever. Smallpox was totally eradicated by 1980, a task that took 11 years, involved thousands of health workers around the world, and cost $300 million dollars.

Polio (poliomyelitis) was virtually eliminated in western industrial societies in the 1950s and 1960s and malaria was almost conquered worldwide in 1961. When the U.S. withdrew support (the main source of funding) from the eradication program, a resurgence occurred. Millions of people now contract malaria each year, with the greatest concentration in central Africa where the majority of children are infected (6).

There is growing evidence that several other declining diseases, including tuberculosis, hepatitis, and cholera, are reversing themselves, especially where forced migration, overcrowding, pollution, and drug abuse have lowered sanitary standards, increased malnutrition, and weakened immune systems. Even though curable, tuberculosis is now increasingly rapidly, especially in Asia and Africa, and reportedly 2.7 million people died from it in 1993 (7).

Equally discouraging is the outbreak and spread of new epidemic diseases, such as AIDs (acquired immunodeficiency syndrome), hanta virus, and ebola. Some authorities believe these diseases previously existed in remote areas, were carried by local animals and insects, and then spread to humans who occupied these areas, often due to the press of population growth or for natural resource exploitation.

Laurie Garrett, in her well-researched and absorbing (if depressing) book, *The Coming Plague: Newly Emerging Diseases in a World out of Balance* (1995), concludes that the remarkable growth of human populations worldwide and their "voracious appetite for planetary dominance and resource consumption" has upset the balance of "every measurable biological and chemical system on earth." She reviews the effects of toxic chemicals, the destruction of ecosystems, species extinctions, and other environmental degradation discussed in Chapter 5 of this book. Garrett thinks we should be more concerned about infectious disease outbreaks but are ill-equipped to anticipate and to manage them. She cites biologist Rita Colwell's estimate that we have characterized less than four percent of 5000 known species of viruses and about 2000 of the 300,000 or more bacteria thought to exist (8). These microorganisms are everywhere in our environment, including air, water, soils, plants, land animals, sea life, and solid wastes.

In the United States, the delay in responding to HIV (human immune deficiency virus) led to the rapid infection of 1.5 million people by 1993 and now costs us $12 billion annually for research, education, and treatment costs in addition to the toll on human lives. The industrial nations are in a financial position to respond to these diseases, but many developing nations are not. She notes that some countries cannot afford more than $2 or $3 per capita annually for public health care, even though the barest minimum health care "package" costs at least $8. For comparison, the amount spent per capita for legal drugs alone in 1990 varied from $412 in Japan and $191 in the U.S. to $28 in Mexico, $4 in Kenya, and $2 in Bangladesh and Mozambique (9).

Even when needed drugs are available, they are not always used. Garrett reports that Merck Drug Company developed a drug to treat a waterborne parasite that causes blindness. In a remarkable humanitarian gesture, Merck donated it to the World Health Organization, which in turn made it available to affected nations where about 120 million people were at risk from the disease. Five years later only three million had received treatment due to inadequate transportation, military coups, local corruption, lack of primary care facilities, and other problems. These countries lack national health care programs and adequate local medical facilities.

Garrett goes on to describe numerous locally serious epidemics of often fatal diseases, such as Bolivian hemorrhagic fever, Seoul hantaan disease, Marburg virus, Brazilian meningitis, lassa fever, swine flu, legionnaires disease, ebola, staphylococcus, and salmonella. The situation is especially critical when bacteria such as tuberculosis and staphylococcus become resistant to antibiotics, often because patients fail to take the drug long enough to cure the disease and the surviving bacteria develop immunity.

Vastly increased air traffic is another reason for the rapid spread of diseases, especially those with a long incubation period. In 1950, according to the International Air Transport Association, 17 million people traveled by air within the United States and two million took international flights. By 1990, this had escalated to 424 million domestic and 280 million international travelers. Unlike ships and trains, airplanes swiftly transport people great distances without an opportunity to discover and quarantine diseased individuals.

Effects of Toxic Chemicals

Rachel Carson and many subsequent researchers have perceived a connection between the expanding array of chemicals we add to our air, water, soil, and processed foods, and the increased incidence of cancer and other physical and psychological disorders that are otherwise difficult to explain. In the past few decades, many illnesses and deaths have been attributed to long-term exposure to coal dust, asbestos fibers, lead paint, mercury and PCB (polychlorinated biphenyl) poisoning, nicotine, radioactivity, radon, formaldehyde, sulfur dioxide, DDT, and other pesticides.

New evidence suggests that industrial chemicals and pesticides affect hormonal functions in humans and animals, interfering with reproduction and normal development (10). Apparently combinations of chemicals often have a more powerful hormonal effect than individual chemicals initially tested in laboratories (11).

Global Environmental Change

Many species of plants and animals are adapted to fairly specific environmental conditions and vary widely in their need for sunlight, nutrients, rainfall, and period of growth. Humans are no exception. The first humans apparently were adapted to a tropical climate, obviating the need for furs or feathers or the ability to hibernate. Successive inventions, such as clothing, shelter, controlled fires, irrigation, food preservation, and air conditioning enabled them to extend their range over most of the world. Nevertheless, there are probable health risks for humans as well as other animals and plants if climatic trends such as global warming and the thinning of the ozone layer continue (Ch. 5), and both the World Health Organization and the Environmental Protection Agency have expressed concerns.

An average rise in global temperature of 3-4 degrees Fahrenheit (2 degrees C.) within the next century, with warming unevenly distributed, could greatly increase the number of days above 100 or 110 degrees (38-43 C.) in the temperate zones. Some possible health effects in the United States could be in-migration of insect pests from tropical areas and the diseases they carry, and an increase in heat-related illnesses, such as strokes and heart attacks.

The stratospheric ozone layer protects humans and other species from excessive ultraviolet B (UVB) radiation. Emissions of chlorofluorocarbons and other gases of human origin are thinning this layer and could trigger an increase in skin cancer, including melanoma with its mortality rate of 25 percent. Other possibilities are more eye cataracts, weakened immune systems, and indirect effects from UVB radiation that affects animals and plants (12).

The Promise of Biotechnology

Biotechnology encompasses a wide range of techniques for causing plants, animals, and microorganisms to develop desired traits or to produce needed substances. Well-known traditional applications of biological principles are selective breeding of plants and livestock to obtain new varieties, and the use of molds, yeasts, and bacteria to make a wide assortment of cheeses, breads, and beverages. More recent procedures include using chemical fertilizers to improve plant nutrition, vaccines to confer immunity of livestock to diseases, introducing natural enemies to control plant pests, and artificial insemination to breed superior livestock.

Emerging Applications

Today agricultural biotechnology has moved well beyond these time-honored methods to transferring genes and embryos from one species to another, using laboratory tissue cultures to produce clones of superior plants, injecting hormones to increase milk production or decrease pork fat, and bioprocess engineering. Gene transfer (recombinant DNA technology) involves identifying a desired trait in an organism, finding a gene that will produce it, and modifying the gene so it can be inserted into a selected cell for reproduction. In this manner, plants can be developed that have higher yields, take less time to mature, have more fiber content, have a longer shelf life, or other qualities. Similarly, animals can be made larger, leaner, more productive, or more disease resistant (13).

Biotech enthusiasts anticipate rapid-growth trees, pest resistent vegetables with less need for chemicals, better tasting foods, lower cholesterol foods, hardier plants, and the ability to feed more people with available resources. Genetic engineering also permits the extraction of genes from plant varieties that are being displaced by human activities and "banking" them for future use. As the new high-yield varieties of rice, corn, wheat, and potatoes gain in popularity, the traditional varieties disappear. In the event that insects, a blight, or another disease destroys the new variety, other plants can be developed from the preserved gene plasm (14).

Environmental Concerns

As with any new stage of technology, some people are apprehensive about possibly adverse social, economic, and biological effects of these experiments and the genetically engineered products that result, wondering if the benefits will outweigh the risks. For example, might some of the hormones used to alter plants and animals also affect people who consume them? What happens if these new organisms or imported "natural enemies" of pests escape into the natural environment and proliferate in the manner of killer bees, bee mites, gypsy moths, and fire ants? What if only large producers of foods, livestock, or commercial timber can afford the new varieties and the technical expertise needed, and small producers cannot compete with their high yields or unique products?

According to a 1987 Office of Technology Assessment survey, biotechnology or "genetic engineering" is a moral issue for about one-fourth of the American public. They question whether scientists should be changing the genetic heritage of plants, animals, and ultimately humans themselves. In general, less-educated people and religious fundamentalists are most likely to regard cross-breeding and genetically altering species as immoral. But other people also oppose some of these practices because they support animal rights or they are genuinely concerned about unforeseen adverse consequences (15).

When biotechnology experiments are federal actions, including university research activities that are federally supported, they are subject to federal environmental laws and regulations. These safeguards include the National Environmental Policy Act (NEPA), and the standards set by the Animal and Plant Health Inspection Service (APHIS), the Environmental Protection Agency (EPA), and USDA's Marketing and Inspection Service (M&IS). Agency research guidelines also apply, including those of the USDA's Office of Agricultural Biotechnology (OAS) and the National Institute of Health (NIH). Products thus created are screened and regulated by the Food and Drug Administration (FDA), the Food Safety Inspection Service (FSIS), and applicable state agencies that regulate nonfederal experiments and products (16).

Part Three

The Search for Solutions

Chapter 8

The Environmental Movement

Archaeologists and historians have painstakingly examined the scattered campsites of prehistoric peoples and the more durable ruins of ancient civilizations to learn about their culture, technology, and social life. Many of these communities and societies were decimated or displaced by climatic changes, such as the last ice age, droughts, flooding, volcanic explosions, and other natural events. During much of this early period, nature's whims usually surpassed human influences in altering the environment, disrupting people's lives, and forcing humans and other species to relocate. But human activities, including ethnic rivalries, conquests, and local depletion of game and plant foods, also forced people to migrate or to change their way of life.

Imagine the impact of the last ice age on humans, animals, and plants. Snow accumulated for millennia, creating vast ice sheets that extended over northern North America as far south as Ohio, over northern Europe and Russia, and into southernmost South America, killing or displacing many forms of life. Ocean levels declined, uncovering numerous land bridges that enabled people to migrate to other islands and continents. As the ice sheets melted, ending about 11,000 years ago, enormous lakes formed and then drained, and oceans gradually returned to their earlier level, flooding some of the inhabited coastal plains.

Throughout history, periodic droughts occurred in parts of the world, including China, central Asia, north Africa, the Mideast, Yucatan, and the U.S. Southwest. Without today's technology for preserving or importing foods, affected peoples either endured widespread famine and fuel shortages or emigrated in search of better conditions. When such calamities were local or regional in scale, migration was often a feasible solution because many areas were still sparsely populated.

Today the forces of nature continue to affect people and other living things but often their effects are substantially modified by human activities. Many scientists and other informed people see a potential for disaster from rapidly increasing human influences on the environment. These include accelerated population growth, uncoordinated technological development, unrestrained use of natural resources, and the resulting demands on the biosphere. Others disagree, saying that our rapidly accumulating body of knowledge, technical skills, and management expertise better equip us each generation to remedy the most important threats posed by nature and human activities.

The Human Role in Nature: Conquer or Coexist?

When the Industrial Revolution proliferated in the 1800s, first in Great Britain and then in the United States, Germany, France, and abroad, the prevailing mood seemed to be that technological progress was inevitable and good, that resources were almost limitless, and that the road to riches was meeting the expanding demands of consumers (Ch. 1). Raw materials, including forests, croplands, minerals, fossil fuels, wildlife and fish, and fresh water, were still abundant in much of the world and new industries were established to "develop" them. By 1870, the race to tap cheap natural resources and to amass private fortunes was well underway in Europe and the Americans, and soon other nations followed (Ch. 4). Global exploitation continues today as multinational corporations and developing nations access once remote natural resources in the tropics, on the tundra, in the oceans, and deep in the earth.

Many people are now convinced that the environment is being severely degraded on a global scale, and that we are not doing enough to reverse destructive trends. At the same time, some countries lack sufficient resources to meet their needs and want to reap some of the benefits that industrial societies take for granted. About a billion people live at a bare subsistence level or lower, millions more could be severely affected by crop failure, and emigration is no longer a viable solution for large numbers of people (Ch. 2, 6). Informed people around the world who share these concerns are environmentalists at heart and many of them support or participate in "green" movements promoting conservation and environmental protection.

Critics of these movements say that many enviromentalist claims about harmful trends are unfounded and others are grossly exaggerated (Ch. 9). They often characterize environmental activists as doomsayers who misrepresent the true condition of humanity and the environment, causing widespread anxiety about our future and unnecessary regulation of the economic sector. They believe that human ingenuity can and will save the day without severely restrictive programs to limit growth and to protect the planet's resources. With today's advanced technology, expanding economic opportunities, improved health care, increased food production, numerous public services, and modern conveniences, people never had it so good and conditions are getting even better.

These opposing views are at the crux of the current environmental movement. Both factions are influential enough to force legislative and regulatory compromises on major environmental issues (Ch. 10).

Early Environmentalists

For more than two centuries, thoughtful scientists and lay persons have engaged in this wide-ranging debate about the future well-being of the human species. In 1798, Thomas Malthus, an English social theorist, created quite a stir when he published *An Essay on the Principles of Population* and stated that poverty and distress cannot be avoided because human population increases faster than our ability to supply sustenance (food). Malthus first concluded that war, famine, and disease were the major limitations on population growth, but later added that moral restraint (self-control) could reduce the birth rate. He argued that overpopulation served to stimulate industry and discourage indolence, but public doles tended to encourage larger families.

In the early 1800s, influential Americans, such as Thomas Jefferson, John James Audubon, and Henry Thoreau, began to see the merits of careful stewardship of lands and resources to prevent some of the abuses that had already occurred. Others did not, judging from their actions.

When the United States acquired the Louisiana Purchase from France in 1803, it doubled the country's size and made vast new resources available. Many citizens and immigrants thought that all of the new lands were open to settlement, despite the presence of numerous Native American tribes. Some thought they had an economic right, patriotic duty, and divine obligation to develop and to civilize these areas. Mining, hunting and trapping, timber harvesting, and cropland development proceeded on a massive scale from east to west along rivers and overland trails, and continued as railroads and roads were developed. Until about a century ago, public criticism of unrestricted population growth, unfettered natural resource development, and needless environmental damage was too weak and fragmented to encourage much legislative and agency concern. Then some scientists, journalists, agency professionals, and lay citizens began to foresee a time when relentless population growth and generations of mounting consumption could severely deplete raw materials and seriously degrade the natural environment. Similar concerns were evident in Europe where the population was larger, the land area smaller, and natural resources more limited. Conservation emerged as a philosophy and practice, and ultimately influenced state, federal, and various private natural resource management programs. Some early leaders in this movement were:

John Muir (1838-1914). An American naturalist, Muir was born in Scotland, educated in Wisconsin, and moved to California in 1868. He developed a broad appreciation for nature through extensive travel, but was especially fascinated by California's natural wonders, especially its giant redwoods and Yosemite Valley. In 1892, Muir and his colleagues founded the Sierra Club, an organization dedicated to natural resource conservation and outdoor recreation. This group, which now has a national membership of about 560,000 in 300 local chapters, helped to ensure the creation of national forests and parks, and now supports environmental protection through education, outdoor activities, political action, and formal litigation.

Theodore Roosevelt (1858-1919). Like Muir, Roosevelt traveled extensively in the United States and was impressed by what he saw. As president from 1901-1909, he was able to protect and conserve the country's natural resources in two ways, by instituting stronger federal controls on business practices and by putting large tracts of relatively undeveloped lands into public ownership.

Roosevelt and his supporters initiated a series of measures to regulate industrial activities and to curb excessive and wasteful natural resource exploitation. These included lawsuits to break up large trusts that interfered with free trade, extended federal regulation of interstate commerce, support for consumer legislation including federal meat inspection, and the creation and expansion of the National Forest System in 1907.

Gifford Pinchot (1865-1946). As the first chief (1898-1910) of what became the U.S. Forest Service, Pinchot promoted forest and wildlife conservation, prudent use of forest products, and agency responsiveness to public needs. He was critical of the waste and destruction of natural resources in the U.S. compared to Europe and advocated sustained timber yields from scientifically managed forests instead of unregulated cutting.

He publicized these themes in his books and numerous speeches. Aldo Leopold (1886-1948). As a Forest Service ecologist, Leopold advocated a "land ethic," under which people regard themselves as part of natural communities and respect nature rather than destroy it. He later taught wildlife management at the University of Wisconsin and, with Robert Marshall, was a founder of the Wilderness Society in 1924. His *Sand County Almanac* is still a popular reference for environmentalists.

H. Fairfield Osborn (1887-1969). A prominent naturalist and chairman of the Conservation Foundation, Osborn wrote *Our Plundered Planet* in 1948 (when the world had 2.3 billion people) to describe what humans had done to the earth and the speed at which they were "destroying their own life sources." He expressed concern about our waste and misuse of natural resources and predicted a grim future for humanity, unless we change our ways. The book became a best seller but Osborn was dismayed when his detractors labeled him a "prophet of doom." He was convinced that the world's population is increasing while its natural resource base is declining and thus edited and published *Our Crowded Planet* in 1962. Audubon Society's William Vogt expressed similar concerns in *Road to Survival*.

The Modern Environmental Movement

Following the industrial expansion of World War II and the postwar recovery, natural resource depletion and air and water pollution were increasingly obvious in much of the world, along with mounting public pressure for remedial action. Symptoms included thick palls of smog over many industrial cities, rising mounds of urban wastes, large areas devastated by economic development and erosion, contaminated streams and lakes, declining wildlife populations, and Jacque Cousteau's lament that the oceans and seas were becoming increasingly polluted.

By the 1960s, many people believed the day of reckoning was close at hand. World population had doubled in 40 years and was expected to double again in another 40. U.S. population had increase 500 percent in less than a century and was growing rapidly due to a "baby boom" and immigration (Ch. 2, 6). Around the world, forests were being cut and land was being converted to agriculture, mining, industrial plants, homes, highways, parking, utility corridors, public services, recreation, landfills, toxic dump sites, and the other needs of growing populations. There was growing public awareness of these trends and a willingness to respond to them.

Several best-selling books of this period described the problems at hand and called for action. In 1962, Rachel Carson's award-winning book, *Silent*

Spring, was published and, with Fairfield's *Our Crowded Planet*, is often given credit for launching the current phase of the environmental movement. Carson, a marine biologist, described the dangers of using millions of pounds of pesticides annually. These products pollute the air, water, and soil, kill countless birds and fish, and endanger human health. She was especially critical of DDT, a toxic agricultural insecticide that Congress banned for U.S. use in 1972 but is still sold abroad.

In 1968, biologist Paul Ehrlich's small book, *The Population Bomb*, predicted severe famines in the decades ahead and called for American leadership in curbing rampant population growth, improving agricultural production worldwide, and reversing environmental deterioration. That year human ecologist Garrett Hardin introduced the principle of "tragedy of the commons" to describe the fate of resources that are free and unmanaged, such as the oceans, open ranges, well water, and the atmosphere. Unlike most people who own private property or visit managed preserves, he thought, those who use the commons feel free to take any amount of resources and to pollute as much as they want. Eventually the combined impacts degrade the commons and all of the users lose.

At this time American college courses dealing with population growth and environmental degradation increased in popularity and dozens of college textbooks and anthologies were written to meet this new demand. Enrollment in the author's own course increased from 19 to 55 in three successive terms. The national environmental movement was publicized during Earth Day 1970, an event that generated widespread citizen involvement continuing to this day. College professors offered marathon lecture sessions and community teach-ins, older students organized demonstrations, and school children picked up litter.

In 1971, ecologist Barry Commoner's book, *The Closing Circle*, summarized what he called the "environmental crisis" brought on by human efforts to conquer nature and to produce wealth and consumer goods. We like to think we are improving the quality of life for humans, he observed, but we are doing the opposite when we degrade the environment. He also noted that population growth is not the sole basis for environmental deterioration. Other key factors are the ability of a nation's technology to avoid waste and pollution, its economic growth to support additional people, and its reduction of poverty and illiteracy so that people will voluntarily limit family size.

In 1972, Dennis Meadows and his associates published *The Limits to Growth*. They used computer models to project current population and economic growth trends and predicted that if these continued, the limits to growth will be reached within 100 years. The most probable consequence of this would be a rather sudden and unavoidable decline in population and industrial capacity. The authors conceded that we still have time to alter these trends and the sooner we begin, the better. Also released in 1972 was *Only One Earth: The Care and Maintenance of a Small Planet,* by political economist Barbara Ward and biologist Rene Dubos. This report was prepared for participants at the United Nations Conference on the Human Environment held in Stockholm in June 1972 but intended for a much wider audience. Compiled with the assistance of 152 people from 58 countries, it discussed the nature of the planet's environment and how humans impact it. By this time the environmental movement was well-established, global in scope, and getting significant public attention and legislative action. In 1980, the U.S. Department of State and the President's Council on Environmental Quality (CEQ) published and released the *Global 2000 Report to the President* (Jimmy Carter), summarizing population growth trends, expanding natural resource consumption, impacts on the environment, and the need for international cooperation to deal with these issues. In 1982, the World Resources Institute was founded in Washington, DC, to conduct research and to promote sustainable development. Its annual report, *World Resources,* provides a variety of current population, energy, agricultural, health, and natural resource data from over 150 countries.

The *Global 2000 Report* stimulated a great deal of discussion and evoked strong rebuttals from pro-development interests who said the grim depictions of resource depletion and environmental decline were excessive and sometimes erroneous. Futurists Julian Simon and Herman Kahn soon edited a book, *The Resourceful Earth: A Response to Global 2000* (1984), that disagreed with the report on many issues and predicted that the world would become a better, cleaner place in the years ahead due to advances in technology and the industrial sector (Ch. 9). In an earlier work, *The Ultimate Resource* (1981), Simon had asserted that there are no real limits to population and resources on earth, that more people each generation is generally beneficial, and that both cost and scarcity of resources decrease as more material is used.

As the environmental movement progressed and research data accumulated, more developments unfolded. More voters got interested in environmental issues and these provided an additional basis for electing or defeating candidates for elective office and for agency appointments. Many federal, regional, and local programs to protect the environment were planned and implemented, including recycling, reducing pollution, preserving wildernesses, and reclaiming abandoned mine sites. New issues, such as ozone depletion, global warming, and the increasing pace of species extinctions added gravity to the environmental dilemma and challenged additional scientists and agency specialists to get involved.

Some people sought a middle ground between the grim scenarios of some environmentalists and the bubbling optimism of euphoric supporters of the status quo (Ch. 9). They thought we should increase our understanding of the problems we face and develop sensible, feasible responses. Perhaps no one epitomizes this middle ground better than biologist and ecologist Barry Commoner, who has been an influential and widely respected participant in the environmental movement since 1970.

In *Making Peace with the Planet* (1990), Commoner explains how humans have transcended their own social world and now exert a strong influence on the structure and dynamics of the natural world. Prior to the last few generations, the planet was regulated chiefly by natural processes, the interaction of the sun, the atmosphere, the bodies of land and water, and the flora and fauna they support. Human interference with natural cycles was limited (Ch. 1).

Modern technology and expanding consumption of natural resources are now changing this relationship, as we increase the globe's temperature by burning fuels, harm life forms with our air and water pollution, relentlessly harvest virgin materials and species, destroy viable ecosystems, and produce huge quantities of toxic and solid wastes. The planet's biosphere has limited flexibility and adapts slowly to changes, while human technology introduces numerous types and massive quantities of substances that are often alien to nature but can interact with and disrupt natural processes. Humans and their technology must be compatible with the natural order for both to survive and prosper. If we are to resolve these conflicts and make peace with the planet, we must first understand what is happening.

The need to act is urgent and the task before us is complex, even intimidating. Past efforts to resolve these problems have had only limited success. But many factors are now in our favor, including the end of the Cold War, advances in technology, better research data, and the increased willingness of nations to work together on mutual problems. We have the ability to modify our productive processes and can make them much more compatible with nature, while still producing the things we need and meeting the needs of millions of people who now live in poverty. This is an enormous undertaking involving industry, agriculture, transportation, and power production, and we must increase public consensus and divert resources to do it. But our technological successes of the past few decades demonstrate what we can do when we put our collective minds to it.

Sustained Public Awareness

Earth Day 1970 and the events that followed greatly increased public awareness of environmental conditions and today a majority of Americans and the people of many other nations support measures to protect the planet, even though they may disagree on the details. When Earth Day 1990 was observed, a New York Times article estimated that 200 million people in 140 nations "from Antarctica to Zimbabwe" took part in grass roots events. Millions in the U.S. participated with demonstrations, exhibits, speeches, seminars, celebrations, recycling drives, cleanup campaigns, and tree planting (1). The 1992 United Nations Conference on Environmental and Development in Rio de Janeiro, Brazil, drew official representatives from 178 nations and hundreds of organizations. Earth Day 1995 was also widely observed.

Several professional opinion survey firms (Gallup, Roper, Harris, and others) have sampled public opinion on environmental issues since 1970. Their findings reveal consistently strong public support for environmental protection, even if it means higher taxes. A Harris international survey conducted in 1988-1989 demonstrated similar concern and support in 15 countries, including some developing nations in Asia, Africa, and Latin America. Concerns included polluted air and water, toxic chemicals, loss of farm and forest lands, and the overuse of pesticides and herbicides (2). In the 1990s, the United States has made additional efforts to protect the environment, although several other nations now seem to surpass us in their willingness to conserve energy, to reduce solid wastes, and to respond to other environmental conditions.

Major nature-oriented societies have generally increased their membership and field activities. These include the National Geographic Society, Audubon Society, National Wildlife Federation, Natural Resources Council, Sierra Club, Nature Conservancy, Defenders of Wildlife, Izaak Walton League, and Wilderness Society. Countless new national, regional, and local environmental groups have been organized to deal with specific issues and their numerous newsletters and magazines keep members informed.

The 1993 *Earth Journal Environmental Almanac* lists 116 such organizations devoted to protecting forests, birds, wildlife, whales, rivers, oceans, fresh water, coral reefs, domestic animals, fish, soils, sand dunes, or unique areas, and ensuring environmental justice, manageable populations, resource conservation, and environmental health in general.

The Rise of Extremism

Growing concerns about population growth, natural resource exploitation, environmental quality, and resulting efforts to protect the environment have spawned two opposing sets of extremist orientations and groups since the 1970s. Both wings have formulated rationales and published books and articles to support their claims and justify their actions. They are also politically active, recognizing that they represent a minority viewpoint and need to influence more people (Ch. 10). Some outsiders fear that the sometimes confrontational activities of these groups jeopardize efforts of mainstream environmental organizations and responsible businesses and industries that look to the future and seek sensible solutions to environmental issues through the democratic process (Ch. 10). Defenders of these groups think that environmental conditions are critical enough to justify drastic action now.

Radical Environmentalism

On the political left wing, well beyond traditional liberals, are radical environmentalists who are alarmed about runaway population growth, expanding industrial activities, the assault on nature, and society's apparent reluctance to reduce these threats (Ch. 10). This is a small but growing movement in the United States, Canada, and abroad, and encompasses fairly diverse segments such as animal rights activists, Earth First!, and other less structured groups. Radical environmentalism is committed to preserving and protecting nature and it advocates direct, effective actions to help to ensure this.

Although their philosophies, objectives and strategies differ, many of these groups agree that human society, as now organized and functioning, is nowhere near sustainable and must be restructured into smaller units with simpler technology. Their options for doing this include education, demonstrations, direct actions against offending industries, revolution, and offering a better alternative when the present order faces ecological collapse. Edward Abbey's 1975 novel, *The Monkey Wrench Gang*, depicted a determined band of ecoteurs (ecological saboteurs) who destroyed developers' equipment in their efforts to discourage exploitation of natural resources.

According to Martin Lewis, four common principles of the movement are: living with nature is preferable to dominating it, decentralization will improve social and ecological health, technological progress is harmful and dehumanizing, and the capitalist market system is destructive and wasteful (3). In this movement, GNP stands for gross national pollution.

Debunking the Greens

Ultraconservatives, with views well to the right of the moderate conservatism of Eisenhower, Nixon, and Nelson Rockefeller, tend to be very critical of the motives, claims, and successes of environmentalists. Many are unconcerned about the present rate of population growth (some find it desirable), argue that most alleged environmental deterioration is grossly exaggerated, seek to weaken or eliminate most government-imposed environmental regulations and restrictions, and would like to see most public property transferred or sold to private interests (Ch. 9-10).

Key Environmental Legislation

In the 1960s and in later amendments, Congress responded to mounting public concerns by passing legislation to protect air and water quality, wild and scenic rivers, and many of the remaining wilderness areas. The National Environmental Policy Act of 1969 was enacted in 1970, the same year the Environmental Protection Agency was established. Subsequent legislation soon followed to protect endangered species, to control the use of pesticides and other toxic substances, to ensure safe drinking water, and to protect historical and archaeological sites. Additional laws regulated national forest and public lands management practices, mining activities, ocean dumping, and occupational safety. Among the most far-reaching laws are:

Wilderness Act of 1964

This act establishes a National Wilderness Preservation System and provides criteria for selecting and protecting designated wilderness areas. By 1992, about 95 million acres had been so designated, 60 percent of it in Alaska and most of the rest in the western states. The Wild and Scenic Rivers Act of 1968 extends similar protection to selected segments of rivers with unique qualities that merit protection from development.

National Environmental Policy Act of 1969 (NEPA)

This is the centerpiece of federal environmental legislation because it encompasses the total human environment with its physical, biological, economic, and social dimensions. NEPA requires all federal agencies to estimate and consider the effects of their proposed actions on the quality of the human environment. They must also involve interested and affected publics in planning and decisionmaking that affects environmental quality.

When a proposed action (a plan, policy, or project) may have a significant effect on the human environment, an environmental impact statement (EIS) is prepared to inform interested publics and to provide (with other relevant data) a basis for a decision. The EIS describes the proposal and reasonable alternatives to it, including no action, estimates their environmental effects, and measures to avoid or mitigate adverse impacts.

NEPA also established the President's Council on Environmental Quality (CEQ) to evaluate and monitor environmental quality, to obtain and publish information on trends, to recommend environmental policies, and to develop regulations (40 CFR, parts 1500-1508) and guidance to implement the act. Scores of states and foreign governments have used this act as a model for their own legislation.

Environmental Protection Agency (EPA), 1970

The EPA is an independent federal regulatory agency established by President Nixon and assigned the leadership role in setting national environmental standards and enforcing federal laws to protect human health and the environment. Thus EPA determines and enforces standards for air and water quality, emissions from pollution sources, solid and hazardous waste disposal, hazardous waste site reclamation, pesticides, radiation protection, acid rain, testing and transporting chemicals, and acceptable noise levels.

Endangered Species Act of 1973

This act sets standards for designation of threatened and endangered plant and animal species by the U.S. Fish and Wildlife Service and provides for their protection, in cooperation with other federal agencies and state and foreign governments. As of 1994, 436 animals and 526 plants were listed as threatened or endangered in the U.S., along with additional foreign species.

Major International Initiatives

In 1972, representatives of 113 nations met to develop the United Nations Environment Programme for international action to protect the global environment. Since then, more international conferences have been held each decade with more nations involved in resolving a growing number of issues. The largest, most publicized, and perhaps most important conference was the 1992 Earth Summit in Rio de Janeiro, Brazil. Many concerned people regard this event as a crucial step in international efforts to acknowledge and to ameliorate the emerging environmental dilemma.

Over 100 heads of state and thousands of environmental scientists and practitioners met to adopt measures to protect the Earth's environment and to encourage sustainable development. Most participants agreed that impacts in one part of the world increasingly affect the rest of the planet and that sustainable development practices are essential if the earth is to continue to support life as we know it. They approved a general plan to resolve the earth's ecological problems, with each nation cooperating in specified ways (4).

Many of the other important international conferences on the environment and related social issues held since 1972 are listed below (in chronological order). Appendix B provides brief descriptions.

United Nations Conference on the Human Environment, 1972
London Dumping Convention, 1972
Convention on International Trade in Endangered Species, 1973
Commercial Whaling Ban, 1982
Montreal Protocol on curtailing production of chlorofluorocarbons (CFCs), 1987
Antarctica Treaty, 1991
United Nations Conference on Environment and Development, 1992
Convention on Biological Diversity, 1992
Convention on Climate Change, 1992
Convention on Desertification, 1994
United Nations Conference on Population and Development, 1994
World Conference on Women, 1995.
Habitat II, U.N. Conference on Human Settlements, 1996.

Chapter 9

Optimism or Despair?

Earth Day 1970 and its aftermath stimulated widespread public and professional interest in population and environmental trends in North America and Europe (Ch. 8). By 1972, hundreds of texts, monographs, articles, television specials, and other sources on these issues had appeared and both schools and other receptive publics were deluged with more information than they could digest. Some new sources seemed hastily compiled and sensational in their approach, overstating the perils facing the next generation. These were soon assailed by critics with quite different expectations of things to come.

When professional and public interest in global socioeconomic and environmental issues was gaining momentum, most dedicated scientists and reform advocates focused their attention on relevant topics within their own field. Sometimes their work was further restricted by the assumptions and research base of their discipline, yielding myopic and uneven results. Interdisciplinary studies followed, integrating major human trends affecting the environment, such as population growth, consumption patterns, technological change, natural resource depletion, environmental deterioration, food production, and the effectiveness of remedial actions.

Claims and Controversy

Many of the best selling environmental books of the 1962-1983 period offered either very stark or quite optimistic scenarios of the decades ahead. Rachel Carson's celebrated book, *Silent Spring* (1962), vividly describes how the air, waters, and land have been contaminated with thousands of toxic substances and related hazards, including deadly pesticides that may kill or harm far more species than the intended pests. Victims include agricultural workers, various birds, fish, livestock, orchards, wild flowers, and insects beneficial to crops. Carson recommended alternative biological methods of control, such as sterilizing male pests, using their natural enemies, releasing otherwise harmless bacteria that target pests, or possibly ultrasonic sound.

In the Prologue to *The Population Bomb* (1968), Paul Ehrlich predicted that hundreds of millions of people would starve to death in the 1970s, despite crash programs to feed them. In the long term, he thought, environmental deterioration could cause even more suffering and death. He got the public's attention and no doubt enouraged remedial programs.

The Club of Rome, a multinational group of businessmen, statesmen, and scientists, enlisted MIT professors Dennis Meadows and Jay Forrester to project the future of civilization based on current world trends. Using computer models to track population, environmental trends, and foreseeable changes, they predicted in *The Limits of Growth* (1972) that the globe's maximum feasible growth would be reached within a century. Unless both birth rates and rising industrial output are brought under control, declines in both population and industrial capacity would follow.

In sharp contrast, economist and futurist Julian Simon argued in *The Ultimate Resource* (1981) and related articles that continuing economic growth is essential to meeting our needs and that there are no real limits to population and resources. The world can use the intelligence and creative potential of more people each generation and there will be natural resources to support them because we will find and produce more, and develop substitutes when needed. Other writers predicted remarkable technological innovations, even migration to other planets, that would enable us to become more prosperous and to ensure high living standards for all of the world's people.

The Continuing Debate

In the 1990s, this debate over appropriate population and natural resource trends and policies continues and new insights keep emerging as more people become involved. The diversity of these perceptions is remarkable, ranging from the bubbling optimism of some pro-growth advocates to the somber pessimism of some environmentalists, each view buttressed with assorted arguments, assumptions, and research data.

Environmentalists

In the U.S., the ranks of environmentalists include many middle-class professionals, a large segment of the scientific community, and a cross-section of other people concerned about various aspects of environmental quality. They support measures intended to reduce rapid population growth, rising consumption, unsustainable agriculture and forestry, air and water pollution, resource depletion, and wildlife extinctions.

The most pessimistic observers believe the battle for an acceptable environment has already been lost because remedial actions are slow in coming and insufficient in scope. Existing trends cannot be reversed soon enough to correct our course, so future food and materials shortages are likely, along with a diminished quality of life. Less dour environmentalists believe that prompt and effective action now can protect the biosphere and quality of life for future generations.

In *The Population Explosion* (1990), Paul and Anne Ehrlich note that world population has grown over 50 percent since The Population Bomb predictions of 1968 and estimate that at least 200 million people have since died of hunger. They describe continuing losses of topsoil, declines in available water, and increased environmental degradation due to pollution and other causes. The Ehrlichs express great concern about exponential population growth, rapid depletion of the world's most accessible resources, and a related decline in the quality of social life. They think population control is crucial to resolving other issues, such as preventing wars, famine, poverty, and biosphere deterioration, and advocate voluntary family size limitation now rather than compulsory restrictions later.

In *Healing the Planet* (1991), the Ehrlichs recount some of the gains from the environmental movement but regret that most of the emphasis has been on reducing symptoms of degradation rather than causes, such as overpopulation, unsustainable economic growth, rising consumption, government inaction, and social and economic inequities. We are not yet winning the war for conservation and sustainable growth because we are still spending the "principal" as well as the "interest" in our zeal to exploit natural resources.

Then too, unexpected surprises keep emerging, such as acid rain, global warming, ozone depletion, and new viral diseases. The Green Revolution provides high yields of grain but also depletes soil and groundwater, and reduces genetic variability. Environmental problems are now global in scope and their solutions require national and international as well as local efforts.

In *Beyond the Limits* (1992), Dennis and Donella Meadows and Jorgen Randers portray their earlier Club of Rome study as a warning of things to come if actions were not taken to avoid predicted outcomes. In the intervening years, there have been numerous efforts to increase energy efficiency, develop new materials, recycle wastes, upgrade farming methods, reduce poverty and disease, and protect the environment. They reiterate their earlier view that we can still achieve more equal opportunity, economic stability, and equitable distribution of material goods if we take appropriate actions soon enough.

Current data from their computer model suggest that present trends are still unsustainable; while gains were made in some sectors, options narrowed in others. We cannot reduce resource consumption, pollution, and poverty if population growth continues unabated. We need more emphasis on long-term goals, sustainable production, social equality, and quality of life. Without sizable reductions in energy and materials consumption, per-capita production of food, energy, and factory goods will decrease.

The Meadows cite three ways to respond to the environmental crisis, (1) deny, disguise, or justify its ill effects, (2) treat symptoms, as by reducing pollution or substituting plastics for metals, and (3) change the structure of the system, as by modifying information, goals, costs, and incentives to encourage positive behavior. The last two are essential.

Perhaps the most ambitious private effort to monitor and evaluate global environmental conditions is the work of the Worldwatch Institute in Washington, DC, founded in 1974 by Lester R. Brown. Every year since 1984, the *State of the World*, now available in 27 languages, has updated readers on significant trends and events, such as population growth, social conditions, food supplies, natural resource consumption, health, climate change, and environmental quality.

The Institute has also issued over a hundred topical papers on conditions and trends, a related book series, and the annual (since 1992) *Vital Signs: The Trends That Are Shaping Our Future*. These publications reveal grave concern about world conditions and trends, tempered by reports on successful efforts to conserve natural resources, to decrease population growth rates, to reduce environmental impacts, and to achieve sustainable economic growth.

It's a Matter of Survival (1991) by Anita Gordon and David Suzuki stems from a Canadian Broadcasting Corporation series on the environment. The 1990s are viewed as a turning point in human development, a time when an ecological crisis is forcing us to reexamine our traditional values. "We are the last generation on Earth that can save the planet," states the Introduction, and we should not be lulled into believing we can resolve the situation with a few alterations in our lifestyles. The first chapter, "Beyond your Worst Nightmare," is a scenario of life 50 years from now as depicted by computer models that project a devastated environment and dismal social conditions if exploitation of the planet's resources increases at present rates. The time to act is now, they say, and this may be the kind of monumental crisis that will unite the peoples of the earth and encourage them to work together.

Bill McKibben, in The *End of Nature* (1989), is disturbed by the level of environmental degradation from human domination of the globe and the difficulty of restoring some degree of equilibrium. In a rambling, highly personalized commentary punctuated with vivid illustrations, he depicts the adverse effects of infusing carbon dioxide, methane, chlorine, and other human-source pollutants into the global atmosphere. The "end of nature" refers to the omnipresence of people, their activities, and their accumulating impacts of the natural environment.

As our demands on nature increase, he writes, we get closer to the threshold beyond which nature is neither predictable nor bountiful enough to supply our needs. It will take great ingenuity, new technology, and genetic engineering to keep finding solutions to the problems we create, and a more humble view of humanity's significance in relation to nature.

The Gaia (Earth Goddess) concept suggests that the earth's biosphere behaves as if it were an organism. It was formulated in the 1970s by James Lovelock and Lynn Margulis and revised by Lovelock in *The Ages of Gaia: A Biography of our Living Earth* (1988). The earth's life forms and the physical environment that supports them are integrated and complementary, and more or less automatically regulate climate, species populations, soil conditions, air quality, and other conditions of the biosphere. New species evolve to fill available niches in the biosphere, generally contribute to the health of the ecosystem, and help to permit the continuance of life.

In short, the biosphere is a self-organizing, self-regulating, adaptive system that functions to sustain conditions suitable for life and has done this for eons of time. If human activities disrupt the natural system, it will eventually adapt but perhaps in ways detrimental to humans, e.g., too slowly. He suggests that just a half billion people on earth would probably have too little impact to seriously affect the biosphere. But many billions of people now and in the future, plus their livestock, pets, chemical pollution, wastes, mining, construction, and deforestation, are collectively a major impact.

Pro-growth Advocates

In the United States today, the proponents of growth include many industrialists, investors, farmers, small business owners, and religious fundamentalists. They often extol the scientific and technological achievements of the past two centuries and maintain that we can and should continue this progress into the future. They also have some support in the scientific community and often use economic theory to bolster their claims, arguing that free market forces unfettered by excessive government intervention can take humanity to new heights and resolve most environmental problems as they emerge.

Religious fundamentalists tend to support economic growth and related environmental trends. Many of them say that the earth was created by God for humans to use, that humans (according to *Genesis*) were given dominion over fish, birds, and every living thing, and that people should multiply and populate the earth. Some think that whatever happens is God's will and that God will protect the faithful. Both economic and religious conservatives are politically active and influential, and some people wear both hats (Ch. 10).

In *Eco-scam: The False Prophets of Ecological Apocalypse*, Ronald Bailey, a television producer and science writer, expresses his alarm at the doomsday orientation of Carson, Ehrlich, Meadows, Brown, and Commoner, "whose books sell by the millions and predict calamities that do not materialize." He believes the last 50 years of civilization have shown solid achievements and show that "there is nothing out there we cannot handle." He compares radical environmentalism to earlier predictions of doom by religious sects, and suggests it is a ploy by egalitarians to attack and weaken corporate capitalism.

Bailey critiques the positions of leading environmentalists and then sets forth the conservative view, marshaling evidence to support it. The remedy for continuing population growth is even more, not less, economic growth to meet people's needs and to allow poverty-stricken nations to raise their living standards. Then they will voluntarily limit their family size, as was done in the industrial nations in the 1930s (Ch. 2). He states that even now the world's food and resources are adequate due to the Green Revolution and improved technology but some people lack the income to buy enough of it or are the victims of corrupt governments that inhibit distribution.

In a chapter entitled "The Depletion Myth," Bailey debunks the findings of the Meadows-Forrest computer model and points out that reserves of nonrenewable resources, such as metals and petroleum, will last much longer than projected (at least another generation or two) and that prices will remain low due to modern mining and distribution techniques. Meanwhile the impending scarcities will motivate industry to mine marginal ores with advanced technology, to use more plastics and other substitutes, and to develop new forms of energy. He quotes economist Gale Johnson to assert that "history clearly confirms that 'no exhaustible resource is essential or irreplaceable.'" Bailey dismisses many of the fears of environmentalists, stating that the mass media find these reports newsworthy and pass them on to the public. He doubts that acid rain is more than an environmental nuisance in the United States, that either global warming from greenhouse gases or ozone depletion will be significant problems (a little warming may even be beneficial), and that a new Clean Air Act was necessary or desirable. He says there has been an increase in snow and ice in Antarctica and its western ice sheet is stable. Bailey has great faith that future developments in biotechnology will save lives, feed the hungry, and protect the environment by replacing less desirable products and practices (Ch. 7).

Bailey is also the editor of an anthology, *The True State of the Planet* (1995), that gives some credit to the "first wave" of environmentalism that began in the 1960s but criticizes the claims of disasters that did not happen. He calls for a "second wave" of environmentalism in which government plays a more minor role and private interests have more autonomy, because people tend to take better care of their own property than they do common property, such as the oceans (or the atmosphere, groundwater, roadsides, etc.). He recommends continued economic progress to provide solutions to unfolding problems. Ten other authors develop complementary chapters on different issues, including population and food supplies, pesticides, global warming, biodiversity, water, the oceans, forests, cancer, air quality, and the coming age of abundance.

In *Environmental Overkill: Whatever Happened to Common Sense?* (1993), Dixy Lee Ray, former governor of Washington, agrees that environmentalists and their allies in politics and the media have gone too far in their depiction of an environmental crisis and in their coercive laws and proposals for saving the planet. She believes there are indeed some serious environmental problems, such as air pollution, but that these can be resolved with intelligent application of science and technology. Ray thinks that some of the environmental issues, such as global warming and ozone depletion, are not yet clearly understood and may be specious, while others, such as endangered species, deforestation, and acid rain are not as serious as claimed. We must understand these alleged threats before we devote needless resources and impose excessive restrictions to remedy them.

In *Population: Opposing Viewpoints* (1995), economist and marketer Julian Simon says "there is an impending shortage of people," a position he has supported for many years. From his economics perspective, if we have to pay more for services, for example, to persons who shine shoes, then there must be a shortage of such people. He also asserts that the past few centuries reveal "an amazing increase in natural resources," since they are now cheaper than they were previously. We have discovered more of them and have improved our technology for producing them cheaply.

In the same publication, Benjamin J. Wattenberg writes "there is no environmental crisis" because the environment is "now healthier than it used to be by far." Pollution has been reduced and most of the preventable deaths in America are due to smoking and alcohol. But people like crises and in the absence of a better one, they can focus on the environment.

In his book, *The Birth Dearth* (1987), Wattenberg argues that nations of the industrial West are having too few children to be their future producers and consumers, soldiers, workers, and parents. As a shrinking fraction of the world's population, these countries ultimately will decline in influence and in their ability to defend their institutions and values in a multicultural world. Having fewer consumers and taxpayers will make it difficult to support western industries and public services, such as social security and medicare.

Management consultant Michael Zey's book *Seizing the Future* (1994) is extremely optimistic about the future, predicting that "the human species is about to burst the boundaries of nature and unleash the power of its technology and human ingenuity, hurtling itself to the next stage of its evolution." We already have the technology to build underground housing, mile-high skyscrapers, and artificial islands to house our people. Cold fusion power is being developed to provide cheap, abundant, pollution-free energy. Biotechnology will produce cheap and abundant food, and technology is emerging to improve the environment, so predicted ecological disasters will not materialize. Life span will increase to 110 years due to an explosion in medical discoveries.

Zey thinks these examples are just part of a unified trend of advancements on the technological, artistic, intellectual, and physical levels that are bringing humans into a new stage of societal development. It is the beginning of the macroindustrial era, a time when humanity will excel in six dimensions, time, space, quality, quantity, scope, and size. In most cases, he says, the technology already exists or is being developed. Look for these things to come: high-speed transportation, new energy sources, colonization and extension of earthlike conditions to other planets, tall structures, elimination of scarcity, robots to do household chores, massive amounts of food produced (plant and animal sources), and a world population that is well fed, clothed, and housed for the first time in history. Developing countries are quickly industrializing and their mostly young and adaptable citizens will soon be better producers and consumers.

Intermediate Views

A third and perhaps growing cluster of scientists and citizens agrees with ardent environmentalists that many conditions need to be remedied, and soon. But they think that some concerns have been overstated and that most problems can still be resolved if federal and state governments, the business sector, and the general public will acknowledge them and work together toward a solution. In addition, environmental protection and reclamation are a source of new jobs and capital investment which can help to offset reduced opportunities in extractive and other industries where workers are displaced by machines, automation, and restricted access to publicly-owned natural resources.

Al Gore, in *Earth in the Balance* (1993), calls for an ecological perspective that perceives humans as a growing force in nature, recognizes the long-term consequences of human activities, and clarifies the difficult choices that lie ahead. He summarizes a broad range of environmental concerns, including global warming, ozone depletion, pollution, deforestation, waste disposal, global food and water supplies, and reductions in wildlife habitat and genetic diversity. His persuasive narrative is peppered with anecdotes and examples, such as the tragedy of pollution in eastern Europe and the far-reaching social and environmental effects of temporary climate changes from major volcanic eruptions.

Gore thinks Ronald Reagan's reluctance to act until we better understand environmental problems like global warming is really a choice to continue reckless behavior and thus face more serious problems in the future. Scientific consensus is increasing and we can change course with considerable assurance, seeking both voluntary and political solutions. The United States has the resources to play a leadership role and citizens must expect their leaders to overcome the influence of people who seek short-term gains at the expense of environmental quality and human health.

The collapse of communism and the desire of more nations to follow our example are added incentives to remedy the shortcomings of capitalism, such as needless damage to the environment. Companies typically compute the cost of installing anti-pollution devices, but not the cost of failing to do so. Modern economics needs to calculate and weigh these external costs which consumers must pay, one way or another. Gore likens a dysfunctional civilization to a dysfunctional family; we are addicted to consumption, tend to deny its adverse effects, and suffer spiritually, as our increases in social and psychological pathologies attest.

We must now restore the earth's ecological balance, the "central organizing principle for civilization," writes Gore, and if we agree on our goals, we can achieve this. This is a global problem and getting sufficient consensus among and within nations will be difficult because of tremendous cultural diversity. The post-World War II Marshall Plan to rebuild Europe can serve as a model for a global plan to set goals, coordinate tasks, and guide actions. Suggested strategic goals for this plan are stabilizing population growth, environmental education, using earth-friendly technologies, economic accounting that values ecology, enforceable global agreements, and socially and environmentally sustainable societies.

Aaron Wildavsky, in *But Is It True? A Citizen's Guide to Health and Safety Issues* (1995), states that many of the environmental health and safety issues follow a pattern. A problem, such as toxic wastes or global warming, is identified and then overstated, causing grave concern. This stimulates a response from opponents, especially people who fear losses from remedial measures and deny or belittle the findings. In general, the more we learn about most threats, the less formidable they prove to be, thus reducing the burden of corrective action when needed.

Another remarkably extensive book is the 700-page tome, *A Moment on the Earth: The Coming Age of Environmental Optimism* (1995), by Gregg Easterbrook. The author is clearly aware of a wide range of social and environmental problems and reviews these in the first seven chapters of his book. He seems confident that we are gradually resolving them and will be even more successful in the future due to technological advances and better research data. He cites positive trends in reducing air, water, ocean, and acid rain pollution, toxic wastes, soil loss, radiation exposure, species loss, and waste disposal, and includes numerous examples of these conditions and trends.   Easterbrook examines each of the major alleged threats to the environment in detail, cites its origins and relevant research on the subject, and gives his assessment of the severity of the problem. He also takes a careful look at human and ecological values and their implications for environmental policies. He concludes that most environmental concerns of the past generation have been overdrawn and are not as bad as originally predicted. But these trend reversals are due in large part to prompt human responses to warnings from responsible environmentalists who saw that destructive trends were occurring. The next wave of environmental logic, he believes, will be "ecorealism," an effort to assess the true state of the environment, often misrepresented by both left and right-wing groups. Humans are a part of nature, not a cancerous growth in nature as some radical environmentalists would have us believe. Neither can the earth be recklessly exploited without concern for future generations. Liberals can boast of saving capitalism from itself by 2000, he suggests, by forcing industry to become cleaner and more efficient. We need to move forward sensibly with the health of the planet in mind, and to regard capitalism as a transitional phase toward an ordered society with productive efficiency, equity, and a sense of community. In his view, progress depends on the decline of materialism as we adopt more worthy, attainable goals, and this is feasible.

Barry Commoner's book, *Making Peace with the Planet* (1990), expresses grave concerns about the clash between technology and the planet's ecosphere, and our limited success in resolving these problems. But his analysis is tempered with hope and solutions (summarized in Ch. 8).

Prospects for Developing Nations

Social and economic equity both among and within nations has been a persistent issue in the environmental movement. In *Only One Earth* (1972), Barbara Ward and Rene Dubos pointed out the wide discrepancy between rich nations with high per-capita incomes and consumption, and poor nations with most of the world's people, the highest rates of population growth, and the awesome task of expanding their economies fast enough to meet their expanding needs (Ch. 3).

Gro Harlem Brundtland, in *Our Common Future* (1989), noted that we are now in a historic transitional period in which "the conflict between human activities and environmental constraints is literally exploding." World population will double in the next century and we may need a world economy 5-10 times greater to meet the needs of developing nations that have many poor, hungry, illiterate, and jobless citizens, and often declining per-capita incomes. Similar concerns were expressed at the Earth Summit of 178 nations in Rio de Janeiro in 1992 (Ch. 8 and Appendix B).

Shridath Ramphal, former secretary-general of the British Commonwealth, returns to this theme in *Our Country, the Planet* (1992). The book's Foreword, by Seymour Topping, states that evidence of global warming and ozone depletion is indisputable, and that effective worldwide programs are needed to protect the biosphere. Ramphal emphasizes our need to live in harmony with nature and to modify our behavior toward it. The world is rich in its racial, cultural, religious, and language diversity but this is also a source of stress. Yet humans are bound together by a common humanity, a shared environment and a common future.

Even poor nations have some super rich people, Ramphal concedes, and rich nations have their poor as well as large middle classes that work hard for what they get. But in a global context, middle classes in rich nations are prosperous and people of the poor nations aspire to their standard of living.

Ramphal thinks that countries are not equally responsible for the world's environmental problems and cannot equally contribute to their solution. Industrial countries should make concessions, such as reduced energy consumption, to enable poor countries to make gains, thereby keeping global emissions low. Rather than handouts, developing countries need trade policies that permit economic growth and make them stronger and more able to pay their debts without exporting all of their natural resources at extremely low prices.

Postscript

Many 1970s predictions of environmental disaster proved to be overstated because initial data were faulty, important variables were overlooked, and subsequent events modified outcomes. The prophets who made these grim forecasts believe their early warnings induced changes which have since reduced the degree of environmental decline. Most of them have adjusted their projections to reflect newer data and continue to play influential roles in their fields.

Most independent researchers now agree that even though some key aspects of explosive population growth and escalating demands on the globe's finite resources are not fully understood, there are urgent problems that require prompt reasoned responses. Hence, we need to be better informed, to play a more forceful role in identifying and resolving problems, and to seek increased international consensus and cooperation when issues are global in scope. The health and well-being of future generations depends in part on our ability to conserve natural resources, to contain population growth, and to find effective solutions to the challenges to we face.

But controversy continues unabated. Some very vocal people, many representing special interests, freely disguise opinions about environmental conditions and trends as facts, often with little scientific basis. We see this on talk shows, in product ads, in letters to editors, and in chats with acquaintances. At the same time, responsible scientists and public-interest groups, scattered press reports on global environmental trends, and greater emphasis on environmental education in our schools have significantly increased public and political recognition of the scope and severity of the challenges before us.

We citizens are in the driver's seat and must work toward positive solutions to social and environmental problems. If we don't act prudently, someone else will act selfishly. We must keep abreast of developments and decide whether the people we hear are making valid claims, are protecting their own interests, are sharing group biases, or are simply uninformed. When we wonder what is in store for the world in another generation or two, we should remember this important principle: It's not what we have the potential to do that is critical, it's what we are willing and able to do.

Chapter 10

The Politics of Reform

Many of today's social and environmental problems are global in scope and unprecedented in magnitude. In the 1950s most of the world's people gave little thought to mushrooming population growth, global pollution and warming, large-scale "invasions" of illegal immigrants, widespread abuse of illegal drugs, running out of oil, wildlife extinction, the destruction of rain forests, a resurgence of epidemic diseases, and international terrorism. Either the problem didn't exist, or was local in scope, or was not widely recognized.

Barriers to Environmental Action

Responding to so many emerging global trends is a formidable task in a world divided by religious, ethnic, and ideological differences and by great extremes in income, natural wealth, and technology. A nation of immigrants, the United States is also racially, ethnically, and socially diverse. Personal circumstances often channel each person's perceptions of shared problems, mask their importance, and impede the search for practical solutions.

There are also unifying influences in the modern world, such as rapid transportation and communication, thousands of multinational corporations, dramatic increases in trade, and 50 years of international aid programs, multinational congresses, cooperative agreements, cultural exchanges, and team competitions (Ch. 1, 7).

This chapter explores the U.S. role in seeking political solutions to environmental problems that are mostly social in origin, often global in scale, and a source of continuing controversy. A central issue is the appropriate federal role. Should the U.S., a major polluter and a nation with enormous human and natural resources, take the lead and work with others in resolving global issues? Or should it focus its energies on domestic problems and let other nations and the free market work things out?

Disagreement among Specialists

Many scientists, agency specialists, and others believe that U.S. efforts to resolve critical environmental problems, while sometimes very helpful, are also fragmentary and insufficient. Existing programs reduce the amount of environmental depletion and degradation that would otherwise occur but often fall short of reversing destructive trends and neglect important needs (Ch. 5, 9). They favor stronger, better coordinated measures to reduce escalating environmental impacts from rapidly growing populations and rising consumption.

Critics of this outlook cite the gains under existing environmental programs, stress the need for continuing economic expansion to meet human needs, and claim there are no resource shortages. They trust the potential of new technology to resolve environmental problems and call for a reduction in restrictions that interfere with property rights, needed economic growth, and birth rates (Ch. 9).

Meanwhile, the type and extent of scientific research is strongly influenced by those who supply the grants. While large sums are available for defense, medical, industrial technology, and consumer product research, less money is devoted to basic science, environmental trends, averting social problems, and planning to meet the needs of larger populations. A more equitable balance might dispel some of the indecision about the current state of society and the environment.

Role of Religion and Ideology

It is humbling to reflect that the religious, ethnic, and ideological orientations of any nation or group are the end product of centuries of wars and other social and economic upheavals. More specifically, what each of us believes and practices, whether it is capitalism, communism, Protestantism, Islam, or racism, is the result of a long and tenuous of chain human events. We and our ancestors before us were cast in a mold formed by age-old traditions and superstitions, religious reform movements, forced conversions, charismatic leaders, wars won or lost, international migration, slavery and colonialism, family and neighborhood, formal education, and modern scientific knowledge.

In other words, but for the quirks of geography, history, and genetics, any one of us could have been born somewhere else (if born at all) in a very different social world. History and prehistory are replete with examples of languages, religions, and entire cultures being displaced by some combination of events. Realizing how tenuously we acquire our own outlook and biases may help us to understand and consider how others view the world.

In very traditional societies, religion is often the dominant institution, channeling and restricting people's perceptions and activities. In widely literate secular societies, one's religious and ethnic heritage is modified by formal education, reading interests, choice of friends and spouse, capacity for creative thinking, and other influences.

Political and economic ideologies, such as capitalism, communism, democracy, and fascism, are acquired in much the same way. People adopt mental frameworks and biases which define their world, restrict their experiences, and channel their thinking. To keep their peace of mind, some people even place certain subjects "off limits" from critical analysis.

Other Traditional Values

The private property ethic and the priority of many individual rights over group rights are well institutionalized in America. Some people think they should be free to do as they please with their own property and perhaps even with public lands, despite adverse impacts on other people or the environment. We see examples of this outlook in local neighborhoods where some people fail to maintain their premises and in the misuse of national forests by visitors and lessees. At the corporate and government levels we see it in the reckless use of resources, illegal waste dumping, profiteering by firms that supply essential products, and a very inequitable distribution of income and benefits among employees in some industries.

We also see a strong emphasis on short-term rather than long-range plans and strategies in both the public and corporate sectors. Managers are hired to get results and often receive more bonuses and recognition for immediate gains than for ensuring that their firm or agency will have enough resources and productive efficiency to prosper a generation from now. Note also the relative absence of specific, achievable national goals in the U.S. and the lack of a federal planning board to identify needs and formulate guidelines. Other industrial nations and many U.S. counties and cities have planning teams that routinely assess conditions, recommend goals, and suggest strategies for attaining them.

National Character

Anthropologists coined the phrase "national character" to denote widely-shared behavior patterns within societies, including attitudes toward nature. Some peoples see themselves as victims of nature (droughts, floods, storms) and are grimly resigned to their fate. In contrast, Americans and Europeans persistently seek to dominate nature through expanding technology and development, with vast engineering projects, massive use of resources, and resulting environmental deterioration. Other societies seek harmony with nature, taking what they need without needlessly disrupting natural systems and processes. Many cultural anthropologists and ecologists share the logic of this third perspective.

Some of the most persistent American values were forged in an era of sparse population and seemingly inexhaustible resources (Ch. 1, 4). Progress, in the public mind, is more often equated with economic expansion, material conveniences, and technological marvels than with other worthy objectives, such as optimal health, reduction of poverty, greater literacy, reduced crime and drug abuse, and cleaner air and water. Hence Americans excel in the former and often lag behind other progressive nations in the latter (Ch. 3).

Cultural Inertia

Many Americans view their country as a safe haven, removed from world population pressures, natural resource scarcities, and major civil strife. Perhaps this is one reason why relatively few citizens are active in political parties and the majority fail to vote in almost all elections. Common Cause, a public interest group, notes that only one in a hundred citizens contributes $200 or more to political campaigns each election cycle, so big contributors are very influential.

Many people are unaware that the nation's population has doubled since 1940 and continues to grow briskly because of both natural increase (around 0.6 percent annually) and legal and illegal immigration (roughly 0.5 percent). Or that in one generation we have evolved from a major petroleum exporter to a major importer and now import increasing quantities of other vital raw materials as well.

There are also well-informed people who do know these things and realize that today's decisions and actions are increasingly important to the welfare of future generations, both in the United States and globally. But they sometimes differ widely on how we should respond.

Liberal versus Conservative Policies

Most political liberals in American politics today are Democrats with some degree of concern about rapid population growth and environmental degradation, and the desire to do more about them. Political conservatives, usually Republicans in both federal and state constituencies, tend to support the status quo as they see it. A large minority of the voters perceive themselves as moderates and may freely switch parties or remain independent on candidates and issues.

Different points of view are necessary and healthy when they are honestly offered. They help people to clarify situations, give the voter a choice, and reduce the amount of resources wasted on impractical programs. But disagreement is divisive and dysfunctional when people deliberately obscure or falsify conditions for personal gain and delay critically needed reforms. Some of each motive is apparent in the continuing liberal versus conservative dialogue on social and environmental issues.

Conservative or Pro-growth View

The major thrust of conservatism in the U.S. today, is not to conserve resources, but to preserve and to promote traditional free-market economics and to buttress conservative religious beliefs and values. Conservative political and religious factions have merged as a right-wing voting bloc on population and environmental issues. They generally favor strong incentives for economic growth, minimal regulation of industry, reductions in government services, reduced taxation, and defend our very unequal distribution of wealth. Most of them support free trade policies but oppose extensive U.S. involvement in global environmental programs, including family planning.

Political conservatives usually endorse continuing population growth, rising consumption, and freer access to public natural resources. As they see it, population growth stimulates economic expansion and technological innovation, necessary both now and in the foreseeable future. They question the seriousness of much alleged environmental degradation and would like to ease or eliminate restrictions on the use of public and private lands and on polluting the environment. They say that existing environmental regulations increase costs of production, depress business activity, force the closure of some businesses, cause unemployment in extractive industries, and increase the price of consumer goods and services. When "real" environmental problems exist, these can be rectified as needed with existing or emerging technology.

The conservative perspective receives strong support from a combination of major timber and mining corporations, resource-dependent ranchers, business interests, and employees of these groups, who see their profits or jobs threatened when their access to resources is limited. They have advocated "taking" legislation that would require taxpayer compensation of private landowners if compliance with environmental regulations reduces the economic value of their property.

They grant millions of dollars to various conservative "think tanks," institutes that fund research and publications in support of these views. Their well-financed activist groups, such as the Wise-Use Movement and the Western States Public Lands Coalition, strive to enlist public support for removing or reducing restrictions on the use of natural resources, including public reserves, and favor harvesting ancient forests, mining in the national parks and Alaskan game refuge, and weakening the laws protecting endangered species, air and water quality. Historically, these growth trends have hastened social and environmental changes, such as revolutionary new technologies, massive conversion of lands to new uses, rapid exploitation of resources, displacement of families and communities, and new family roles. Yet conservatives also extol traditional values, such as family unity and responsibility, religious participation, upright moral behavior, public piety, school and ceremonial prayer, and the right to own assault weapons.

Liberal Perspective

Liberals feel less constrained by traditional beliefs and practices, believing that humans are capable of improving their lot by identifying problems, evaluating alternative courses of action, and initiating needed reforms. Most seem to agree that population growth, the accelerating drain on resources, and related environmental impacts threaten the welfare of future generations, and that we need to work toward sustainable population growth and prudent economic expansion.

Many liberals also call for a reduction of poverty, a more equitable distribution of resources, better health care, fewer unwanted and neglected children, and more educational and career opportunities for people who want to get ahead, despite handicaps. Most of them support environmental research and restoration programs, family planning, the right to early abortion of unwanted pregnancies, and government aid to such programs, both in the U.S. and abroad. They feel that governments as well as the private sector must be involved to ensure that social and environmental goals will be met and are willing to pay taxes for effective programs.

There are interesting exceptions to these generalizations, including conservative environmentalists, who look beyond their immediate needs and plan for the future. Some are very critical of the reap, ruin, and run philosophy that has been so prevalent in American history. There are also liberals who think that too much regulation of the economy or state and local governments will stifle people's initiative and creative potential, and that long-term financial aid, whether to nations or families, can encourage dependency. Some oppose abortion except in unusual circumstances and would prefer much more emphasis on preventing conception of unwanted or potentially neglected children.

Environmental Protection: The U.S. Experience

During the past 25 years, liberal versus conservative differences have erupted in controversies over environmental issues affecting various industries, their employees, government agency personnel, and millions of people in other countries with aid and trade ties to the U.S. Federal environmental management during this period illustrates both constructive and dysfunctional controversies. As will be evident, it is often difficult to separate valid claims from distortions and hyperbole.

An Auspicious Beginning

Following World War II, the United States was at the forefront of efforts to improve the quality of life in both developing countries and war-ravaged nations. Both unilaterally and in cooperation with international organzations, the U.S. helped many other nations to upgrade their health standards, to improve their technology, and to develop more self-sufficient economies (Ch. 7). Twenty years later, the U.S. was a leader in the emerging global environmental movement (Ch. 8).

Until about 1980, these efforts were clearly bipartisan and broad in scope. During the Democratic Johnson administration (1963-1969), various federal and state laws were enacted to protect the environment and in 1970 the Council on Environmental Quality (CEQ) and Environmental Protection Agency (EPA) were established to oversee the process (Ch. 8).

Republican Presidents Richard Nixon and Gerald Ford (1969-1977), continued to support the environment. The EPA, headed by William D. Ruckelshaus, set and enforced the new environmental quality standards.

The CEQ, with a staff of 50, fulfilled its intended role of monitoring environmental trends and advising the President on a variety of issues.

The Democratic Carter administration (1977-1981) emphasized energy conservation and issued the 1980 Global 2000 Report which, despite alleged shortcomings, focused public and agency attention on the deterioration of the environment and potential consequences if present trends continued.

The Reagan Reversal

After 1980, the United States government redefined its role as a world leader in working toward conservation, sustainable development, and protecting the environment. Environmental policies were abruptly redirected during President Reagan's administration (1981-1989) on the assumption that the global crisis had been grossly exaggerated. Federal energy conservation measures were largely abandoned, support for family planning programs abroad was discontinued, and action on problems such as global warming was deferred, pending further study. Ardently pro-development administrators were appointed to leadership positions in EPA, the Department of Interior, and other natural resource agencies. Critics described them as "foxes guarding the chicken coop." The staff of CEQ was cut from 49 to 8. These and other moves opened more public resources to private interests, including national forests, grasslands, deserts, and wildlife refuges, with their timber stands, and deposits of minerals and fossil fuels. Political influence reduced the effectiveness of federal agencies charged with managing and protecting natural resources and resolving environmental problems. For example, the Forest Service was directed to allocate more timber for harvest than its planners and researchers deemed advisable.

Some of the resulting actions enraged earth watchers and the general public, and eventually led to the removal of the most glaringly miscast federal officials through dismissals and resignations. Under the subsequent Bush administration, both the EPA and CEQ were less constrained and more effective. But efforts continued to make more public resources, such as protected wetlands and oil from the Alaskan game refuge, available to private interests.

Clinton-Gore Policies

The 1992 national elections produced mixed results for the environment when Republicans won control of both houses of Congress. Democrat Bill Clinton was elected president, partly because 19 percent of the voters, the majority of them Republicans, backed Ross Perot. Clinton thus lacked a strong public mandate for environmental reforms but did take some steps, including an increased emphasis on conservation policies and restoration of U.S. support for family planning programs overseas. Al Gore, a former senator, a staunch environmentalist, and author of *Earth in the Balance*, was elected vice-president and has been influential on environmental policies. (Ch. 9). During the last year of President Bush's tenure, the United States was a reluctant participant at the Earth Summit in Rio, first threatening not to attend if certain conditions were not met, then refusing to sign the biodiversity treaty and insisting on reduced efforts to stem global warming and deforestation. A year later, President Clinton named 25 leaders of industry, government, labor, environmental, and civic groups to a new Council on Sustainable Development. This group helps to develop U.S. policies on economic growth, job creation, and environmental protection. The President also agreed to abide by Earth Summit agreements to cap pollution emissions at 1990 levels by the year 2000.

President Clinton created an informal office on environmental policy in the White House, coexistent with CEQ, and the latter again lost both staffing and influence. Clinton planned to abolish CEQ legislatively and to add its functions to a new Department of the Environment which would also incorporate the EPA. But more than 30 conservation organizations and several influential members of Congress exerted pressure to retain CEQ as a separate body to oversee the National Environmental Policy Act and to advise the President on environmental matters (1). The President then merged his White House environmental unit with CEQ, thereby increasing its staffing and responsibilities.

The Republican Congress

The election of the 104th Congress in 1994 ended Democratic Party domination, as Republicans claimed 52 of 100 seats in the Senate and 230 of 435 in the House. Many Republicans interpreted this victory as a popular mandate for major changes in environmental legislation and the way it was being implemented. They announced their now-famous 1995 "Contract with America," promising tax cuts, reduced federal spending, and a diminished federal role.

Some of the the proposed tax cuts were controversial because they seemed to focus on the poor, the handicapped, and the elderly. At the same time, Congress was willing to give the peace-time military more funds than requested and unwilling to reduce billions of dollars in "corporate welfare," including below-cost timber sales, irrigation subsidies, low grazing fees, virtually free minerals from public lands, and various other taxpayer concessions to industry (2).

The 1995 "emergency timber salvage" rider, part of a larger bill to reduce the budget, permitted harvesting both dead timber and 270 million board feet of green timber in old-growth forests without the usual environmental restrictions. According to the Pacific Rivers Council, a public interest group, this harvest will cost taxpayers $370 million, including the cost of rehabilitating the environment. Environmental groups have sued to block the sale, while timber companies have sought legal support for the action (3). At this writing, the congressional effort to rescind or severely weaken the Endangered Species Act has not succeeded (4).

In 1994 state elections, Republican candidates won 24 of 36 governorships. In the present political climate of many states, a majority of legislators and some governors regard environmental protection measures as excessive, as based on false or exaggerated premises, and as threatening to economic expansion (Ch. 9). Politicians seek to reduce federal and state environmental restrictions on industries and major private property owners, both of whom contribute heavily to their campaign funds. Some formerly moderate political leaders, who find that long-term goals are increasingly difficult to formulate and to defend, are now preoccupied with poll-watching, assessing the winds of change, appeasing supporters, and getting reelected.

Determined Extremists

Population policy, economic growth, natural resource depletion, and environmental protection needs are four of the major battlefields where the extreme left and right lay siege. The term "extremist" usually applies to both reactionaries (ultraconservatives) and radicals (ultraliberals) at opposite ends of the political spectrum. Their objectives, rationales, and activities are well beyond the scope of the conventional liberal and conservative philosophies and the platforms of the two major political parties.

Common Elements of Extremism

Political, economic, and religious extremists, whether radicals on the left or reactionaries on the right, are strongly convinced that people with either moderate or opposing points of view are wrong, misguided, and perhaps subversive. Extremists have their own world views, often reject compromise solutions to pressing problems, and may use unorthodox, illegal. or even violent means to gain their ends. Radicals, by definition, favor extreme changes in new directions from existing beliefs, values, policies, and practices. Reactionaries exalt traditional, often waning beliefs and policies and seek to uphold them or to restore them if displaced.

Such extreme polarization is encouraged by several factors. Each of us is socialized to accept some pattern of beliefs and values, usually reflecting the dominant influences in our lives. When these influences are extreme, persistent, and reinforce each other, a narrow outlook often results. Many people find extreme viewpoints easier to comprehend and to accept than balanced but more complex perspectives. It is also easier to accept the one-sided views of peers than to try to be objective and risk censure.

People who expect to lose money, jobs, or status because of present or proposed actions may also react in an extreme way because they feel seriously threatened and unable to resolve the situation by other means. We see this with some ranchers, loggers, and industrialists who resent reduced access to resources on public lands. Extreme views attract public and media attention, and those who hold them sometimes have an easier time marketing their books, articles, and lectures than do others with more reasoned and moderate but less sensational views. Fortunes have been made by promoting extreme viewpoints that later failed to pass the tests of time and scrutiny.

Extreme views befuddle the general public and hinder scientists, legislators, and agency personnel who seek to analyze and to resolve environmental issues. Politicians and voters become confused when they are faced with diametrically opposed claims about the merits and faults of some trend, proposal, or program. Biased authors and charismatic speakers can reduce complex problems to simplistic claims with little hard evidence and neutralize months of careful effort to remedy a problem. Most significantly, needed actions may be delayed while a problem intensifies when the people who should be responding are assured that nothing needs to be done.

Strongly-held beliefs and rivalries arouse strong emotions and some extremists resort to lying and misrepresentation, character assassination, property destruction, assault, and even murder to gain their ends or simply to make their point. Extremists may think that lies, distortions, and desperate acts of violence and vandalism are justified, but these can disillusion previously sympathetic people and weaken their support.

Anti-Environmentalists

Reactionaries seek to preserve or restore the status quo, for example, unregulated capitalism, low taxes, broad private property rights, limited role of government, inflexible religious conformity, or notions of racial, ethnic, and male superiority. They fear that their perspectives, often described as the American way or God's will, are being threatened or ignored by others and should be accepted, respected, and even part of public policy. Some members of the religious right who believe that God intended humans to dominate the earth and multiply, say that the church's mission is to save souls, not to rescue the earth or life on it, which may be short-lived anyway (referring to biblical revelation). Many advocate suspending U.S. aid to family planning programs at home and abroad, and outlawing all abortions. Extremists have damaged abortion clinics and assaulted their personnel.

Reactionaries deny the "environmental crisis," claiming that environmental conditions are improving, and that emerging technology and the free market left alone can ensure a better world. Some of their spokespeople enjoy "green bashing," that is, harsh criticism and ridicule of the environmental movement and its supporters, who have been called "eco-wackos" "eco-freaks," and "junk scientists." (5). A prominent radio talk show host says that environmentalists fall into two categories, "socialists and enviro-religious fanatics," both set on crushing the American way of life (6). Right wing extremists have used force or threat of force against environmentalists and agency employees to make their point.

Radical Environmentalists

Radical environmentalists tend to think that "old ways," such as rapid population growth, unbridled capitalism, relentless development, and boundless consumption are unrealistic, unsustainable, and indefensible practices and the main reasons for our current global environmental plight. They believe the time for action is now and the mission is critical, namely, the survival of the planet as a habitable ecosystem.

Their most immediate goal is to prevent further desecration of the environment by active means, including demonstrations, sit-ins, tree hugging, road blockades, sabotaging equipment, and other measures. Many radicals want to formulate new national goals and to restructure social institutions on a smaller scale to meet human and environmental needs (Ch. 8).

Several active networks are said to exist in the United States and other countries as well (7). Foremost is Earth First!, which has as its motto, "no compromise in defense of Mother Earth," has used tree spiking and eco-sabotage to impede logging and mining, and has advocated intentional neglect to deal with overpopulation and starvation. Many radicals believe that humans are far too numerous and threaten biological diversity and the future quality of life on earth (8). Some follow a "deep ecology" philosophy which teaches that all life has value and a right to exist, whether or not it is useful to humans.

Radicals as well as some liberals think the world is already running out of some important natural resources and that stern measures are needed to conserve, to recycle, and to preserve what is left. They fear that we are still losing the war against air, water, and soil pollution, waste disposal, species extinction, and many other excesses of modern life.

Mainstream environmental groups, such as Audubon Society, National Wildlife Association, Natural Resource Defense Council, and Sierra Club, commonly work through the system to achieve their goals, relying on educational programs, magazines and news letters, participation in election campaigns, and legal action. But many radicals think these methods are too slow, often ineffective, and also counterproductive when there are compromises with groups that exploit resources and devastate the environment.

Moving Ahead

Fortunately a growing number of governments, industries, and public interest groups are aware of the long-term consequences of many current activities and trends affecting the environment and are determined to alter our course toward better-informed and more sustainable social and economic development (Ch. 8, 11). This global effort is supported and augmented by hundreds of professional associations and millions of private citizens in the United States and abroad who share a concern about the quality of the environment we leave to future generations.

One prominent example of professionals organized to promote a healthy environment is the Union of Concerned Scientists, based in Cambridge, Massachusetts. This network of thousands of leading scientists and their supporters from many fields of study examines the most serious environmental problems and works toward feasible solutions. The Union issued a "World Scientists' Warning to Humanity" in 1993, highlighting major sources of critical environmental stress, the impact of unrestrained population growth, and the urgent need for remedial action. This was followed by a series of briefing papers on specific problem areas. The group seeks to encourage responsible stewardship of the global environment through technical studies, educational programs, and government policies.

Chapter 11

Bold Actions and Mixed Results

Harnessing Human Potential

Many of us agree that humans have the intelligence, ingenuity, and technological potential to protect the environment, to conserve resources, and to ensure an adequate standard of living for generations to come. If so, the key questions are: Will we assess the situation honestly and act soon enough? Will we use enough of our ingenuity and technology to meet these challenges (Ch. 12)? This is difficult to predict, but clues are evident in what we have done thus far.

Since Earth Day 1970, more people in each nation have become aware of the importance of a healthy environment and many of them have tried to reduce their impacts on it. National, community, and individual efforts to preserve and enhance environmental quality have been creative, timely, and helpful. These include mandated conservation of forests and wetlands, protecting endangered species, reductions in pollution and waste, reclamation of eroded and contaminated soils, restoration of lakes, streams, and bays, increased energy efficiency, voluntary recycling, and family planning. This chapter highlights selected local, regional, national, and international efforts to conserve resources and to protect or restore the environment.

Energy Conservation

The world consumes vast quantities of fossil fuels and is increasingly dependent on them even though fossil fuels are limited in quantity and may be insufficient in a generation or two (Ch. 3-4). In addition, extracting fuels from underground sources and transporting them may degrade other resources, while using or spilling them pollutes our air, water, and soil, and appears to be a major cause of global warming. There are also health effects: the American Lung Association estimates that Americans spend about $100 million annually on health-related costs of air pollution.

These are strong incentives to develop energy sources that are both renewable and less polluting. The United States now imports half of its oil (a main source of its annual trade deficit) and feels obligated to protect its foreign sources. Note how quickly the U.S. and Europe interceded when oil-rich Kuwait was invaded by Iraq, but their reluctance to act during four years of murder and mayhem in Bosnia.

Wind and Solar Energy

Since ancient times, wind power has been used to propel boats, pump water, and activate simple machines. With the advent of electricity, rural Americans used wind chargers to generate current for lights and small appliances. In the United States and Europe today, larger and more efficient wind turbines are being erected in breezy areas to generate electrical power. Although these turbines are noisy, visually offensive to some people, and sometimes kill birds, they produce pollution-free power a small fraction of croplands, pastures, or deserts. Solar power plants are most practical in very sunny areas and sometimes require considerable land surface, just as conventional power sources do if transmission corridors, coal mines, dam sites, and power plants are considered. But small solar systems can be placed directly above or adjacent to the facility they are serving, eliminating the need for a power plant and transmission lines. Once solar systems are manufactured and installed, they have few environmental impacts.

Hydropower

Some areas of the United States and Canada, such as the Pacific Northwest and Tennessee, and parts of many other countries enjoy low-cost hydropower. This is a relatively nonpolluting energy source once the installation is finished, but also has several environmental costs. The massive dams needed to generate power create large lakes that submerge farmlands and wildlife habitat, interfere with fish and game migration, and increase sedimentation and eutrophication by slowing the flow of streams. For these reasons, and because dams now exist on most acceptable sites, further expansion of hydropower in the U.S. will be limited. Many large projects now planned or underway in China, Africa, India, Canada, and elsewhere have become very controversial and some have been postponed or abandoned.

Nuclear Energy

In the 1950s, nuclear fission power emerged as a very promising energy source and by 1993, over 400 nuclear power plants in 27 countries produced about six percent of the world's commercial energy. The United States has the most (110), followed by France, Japan, Great Britain, Russia, Germany, and Canada, but expansion of this industry has slowed. The 1979 nuclear accident at Three Mile Island, Pennsylvania, and the 1986 disaster at Chernobyl, Russia, increased public concern about the safety of nuclear power production. Related problems are thermal pollution (warming) of streams and lakes from water used to cool the reactors, the safe disposal of accumulating nuclear wastes, and the expense and environmental costs of mining nuclear fuel (Ch. 5). Scientists are now attempting to develop controlled nuclear fusion as an energy source. Fuel for fusion is much easier to obtain, less pollution would occur, and energy could be plentiful and inexpensive.

Geothermal and Biomass Sources

In some areas, geothermal energy can be developed by tapping subsurface hot water or steam and using it to heat buildings or to generate electricity. Deposits of extremely hot water are limited in extent, widely scattered, and often remote from cities, ruling this out at a major energy source for most nations. This technology can cause pollution if toxic minerals or gases are released into the surface environment.

Biomass power is obtained by burning plant materials, including wastes from timber production, farming activities, and combustible municipal garbage. Biomass is thus a renewable resource and a small but growing source of commercial energy. Croplands or forest lands are needed to grow biomass materials when discarded wastes are insufficient and these power plants produce soot, ash, and gases, pollutants that can be reduced with control devices.

Innovations in Transportation

Between 1974 and 1986, the United States made remarkable progress in conserving oil when the auto industry met stringent federal standards and doubled the fuel efficiency of new motor vehicles. Impressive reductions in energy use were also achieved through better home insulation and more efficient electrical appliances. Little improvement in auto mileage has occurred since 1988 due to reduced emphasis on this program. Imported oil declined in price, federal measures were less stringent during the Reagan and Bush administrations, and the national sense of urgency about fuel shortages has diminished.

California is spurring auto manufacturers to develop new types of fuels by requiring that two percent of the cars for sale in 1998 emit no pollution. Electric cars large enough to carry two or more passengers on local trips are already available, but are expensive and need frequent recharging. Major auto firms are marketing improved versions in 1996 and 1997. These autos depend on commercial power sources, which in turn burn fossil fuels, but are said to offer net reductions in hydrocarbon pollution of up to 90 percent. Invention of a smaller, lighter, less costly, and more efficient battery would make this approach much more feasible.

Vehicles propelled by hydrogen fuel cells are now being developed and tested. In many ways hydrogen is an ideal fuel because cars using it would make little noise, produce no pollution, and have a longer range than electric cars. Like gasoline, hydrogen is explosive, but it can be produced using nonrenewable energy sources, such as wind power. Much remains to be done to make this mode of transport competitive with gasoline-powered autos, but long-term prospects are promising.

Propane is already used to propel some buses and delivery vehicles and these pollute less than conventional gasoline and diesel models. A new "superclean" fuel has been developed by a Reno inventor and is now being tested before it can be commercially available. Developers claim that this blend of naphtha and water can power most internal combustion engines without damage, will emit only about 40 percent of the pollutants of gasoline, and could be competitive in price.

Local Initiatives

Some of the most encouraging signs of environmental reform are the thousands of regional and local efforts to halt deterioration and to restore the quality of degraded sites throughout the world. These range from large-scale efforts, such as Nature Conservancy's purchase and management of hundreds of unique ecosystems in several countries, to cities and industries that strive to reduce air and water pollution, to individual families that remove trash from a creek or pick up highway litter. These are a few U.S. examples.

Connecticut River

In the 1970s, the 410-mile long Connecticut River was cynically dubbed the "most beautifully landscaped sewer in America" because its water was polluted by two million residents and many industries in its drainage. Since the imposition of EPA water quality standards and the efforts of determined organizations, one can now swim, boat, and fish in the river (1). A 76,000 acre wildlife refuge is being established there, much of it on land used for other purposes as well.

Humboldt Bay

The people of Arcata, California, were disturbed that local waste water was going directly into the Pacific. They solved the problem by developing both fresh-water and salt-water wetlands on the shores of the bay. Waste water laden with nitrogen and phosphorus is initially treated in a sewage plant and ponds, and then flows into the marshes where bacteria purify it. These marshes also double as a wildlife sanctuary (2).

Sacramento Utilities

Sacramento's Municipal Utility District (SMUD) in central California strives to reduce energy consumption and vehicle exhaust with an electric trolley and bus network, and other cities are considering similar ventures. SMUD also produces energy from wind power, encourages energy efficiency in building construction and daily consumption, and is developing ways to use solar power. Other states, from the Sunbelt to the Canadian border, also see a future in solar power. Several countries, including Mexico, India, and Japan, are expanding its use in rural areas that lack electricity.

Lake Erie

Two decades ago Lake Erie, the world's 12th largest lake, was too polluted from sewage and industrial wastes for swimming and commercial or recreational fishing. When Cleveland's Cuyahoga River actually burst into flame, the U.S. and Canada acted to reduce the influx of phosphorus and other wastes, resulting in cleaner, clearer water and a resurgence of wildlife and fish. But now zebra mussels from Europe have invaded the lake, with uncertain consequences.

Waste Disposal Quandary

Traditionally, cities dumped their wastes on isolated sites, such as hillsides, dry ravines, bogs, or abandoned quarries. Needy people sought recyclable and reusable items, animals fed on wastes, and the areas were periodically burned to eliminate combustible materials. As populations grew and the volume of disposable items increased even faster, this method proved impractical for urban centers. Space for dump sites was scarce, and air, water, and soil pollution from these dumps was intolerable. Local citizens were also offended by their appearance and odors, and opposed new sites. Alternatives included land fills, incineration, recycling, composting, ocean dumping, and paying other countries to accept wastes.

Most land fills are rural sites where wastes are deposited and periodically covered with earth, reducing both offensive odors and sights. However, seepage of toxic chemicals into groundwater remains a problem and many populous regions have or soon will run out of acceptable sites. One 3000-acre site on New York's Staten Island rivals Egypt's Great Pyramid of Khufu in size. Many people prefer incineration to landfills because it greatly reduces the volume of waste. Modern incinerators use high temperatures to minimize the ash, collect some of the emissions, and recover a portion of the energy used to run them. Yet they produce tons of sometimes toxic ash and airborne gases that pollute the area and may endanger living things.

Recycling reusable metals is a time-honored practice evident in auto wrecking companies and junkyards from coast to coast. Until recently, only a fraction of cheaper metals and other substances was recovered because raw materials were abundant and virgin metals were preferred. Today recycling has expanded to include glass, paper, cans, and plastics, with curb pickup service in many cities. Recycling is a very prudent method of waste processing because it reduces the volume of waste, extends the life of existing ores, and requires less energy than the production of new metals.

But the majority of usable waste in the U.S. is still not recycled, mainly due to the cost of collecting and processing these materials and the limited market for them. European countries and Japan are doing much better (Ch. 5).

Composting, or converting manure, other agricultural wastes, and yard debris into rich earth for gardens and fields, has a long history of use around the globe and is gaining popularity in the U.S. and Europe. Some commercial firms now routinely use organic materials to make soils and additives, and some cities collect leaves and grass clippings to compost for use in their own parks and gardens.

Many coastal cities and ships at sea once dumped their trash in the ocean, causing water contamination, harming marine life, and increasing beach debris. This practice was prohibited by federal law in 1972 and by international agreement in 1988. A few nations found other countries willing to take some of their wastes but, aside from being morally questionable, this simply transferred the problem to another area where waste treatment might be less effective. In the 1980s, the garbage barge "Mobro 4000," with over 3000 tons of New York City garbage, made world headlines when it was tugged 6000 miles in pursuit of a dumping site and finally returned to New York to be incinerated and deposited in the landfill that first rejected it (3).

None of these methods of waste disposal is without problems. Even composting sometimes produces offensive odors and its products may contain toxic substances if these were in the original wastes. The most effective solution is reducing the sources of waste through products that last longer, require fewer materials, use less packaging, and are reusable or biodegradable when feasible. Germany recently passed a law preventing excessive packaging of products, a move expected to reduce trash by millions of tons annually.

Effective Family Planning

Voluntary family planning in Europe is effective in all social classes and most ethnic groups, and the families of four or more children are rare. Largely Roman Catholic countries, including Spain, France, Belgium, Italy, and Austria, resemble neighboring Protestant countries in their low growth rates, indicating that religious affiliation need not dictate family size. The average Italian family now has 1.3 children, as few as any nation on earth.

160   *The Environmental Dilemma: Optimism or Despair?*

Family planning is also widespread in other industrial nations of the world where birth rates usually are higher than in Europe but well below the world average. Some of them have much more land area per capita than most European nations and immigration also increases their growth rate. Due to low birth rates, Western industrial nations are now a rapidly declining segment of world population, falling from 31 percent in 1960 to an expected 14 percent in 2025.

During the past three decades, government-supported family planning programs have been established in most developing nations, often with assistance from the United Nations and private groups concerned about global overpopulation. During 1951-1965, national programs were activated in India, Pakistan, South Korea, Fiji, China, and Egypt, and others soon followed. South Korea, Taiwan, China, Thailand, Mauritius, Singapore, Fiji, and Sri Lanka have substantially reduced their growth rates. The success of these programs varies widely by country, but together they have eased the rate of population increase in each major world region except Africa and the Mideast (Ch. 2).

Results of Environmental Legislation

The national environmental movement gained strength during the Lyndon Johnson and Richard Nixon administrations (1963-1974) when laws were enacted and other actions were taken to protect the environment (Ch. 10). Major new legislation addressed federal environmental policy, air and water quality, toxic wastes, endangered species, and the protection of wildernesses. These acts were later amended to strengthen or refocus them and complementary laws were passed in the states.

Most critics of federal mandates concede that environmental quality in the United States has improved significantly under these measures and sometimes use this as an argument for reducing restrictions on industries. In addition, various international agreements have been ratified to deal with global problems, such as ocean dumping, the impending extinction of whales and other species, ozone depletion, and global warming (Ch. 8). Some of the results of these measures are summarized below.

Clean Air Act of 1963 (amended)

In 1952, 4000 London residents died from respiratory failure attributed to many days of dense smog. The city promptly switched its coal-fired heating plants to natural gas and oil, making the air clearer and less toxic.

Today the air is also cleaner over many other cities in the United States and northwestern Europe due in large part to federally mandated standards for unleaded gasoline, auto mileage, industrial emissions, and safety procedures during local pollution alerts. The auto industry opposed these stipulations for new vehicles as arbitrary and infeasible yet when the government stood firm, its demands were soon fulfilled. Antipollution technology successfully reduced factory emissions and numerous local programs now monitor and control smoke from coal, wood, and refuse burning.

Wilderness Act of 1964

According to the Wilderness Society and other sources, the U.S. has lost over 90 percent of its original old-growth forests in the 48 coterminous states to logging, agriculture, industry, and other uses, leaving about 4.7 million acres, most of it in the Pacific Northwest (Ch. 4). About 30 percent of this remaining area is now protected in designated wildernesses and most of the rest is in scattered tracts on public lands, some of it subject to logging when it is politically feasible.

Formally designated wildernesses are closed to encroachments of modern life, such as logging, mining, roads, motor vehicles, and cabins. Scientists, recreationists, and nature lovers value these areas, which often contain exceptional scenic beauty, relatively intact ecosystems, native plants and animals with increasingly rare gene pools, clean water in streams that originate there, opportunities to study and photograph nature, unsurpassed trail hiking, and freedom from the pressures of modern life.

National Environmental Policy Act of 1969 (NEPA)

The Preamble to NEPA declares a national policy to "promote efforts which will prevent or eliminate damage to the environment and biosphere and stimulate the health and welfare of man." Humans and nature are to "exist in productive harmony and meet the needs of both present and future generations." Under this act, if any proposed federal agency action could impact environmental quality, before a decision can be made the agency must do an interdisciplinary evaluation of the nature and extent of the impacts of the action and reasonable alternative to it, including no action (Ch. 8). This law applies to agency plans, policy changes, and permitted projects such as mining, logging, ski resorts, or utilities corridors on public lands.

When a proposed action's effect on environmental quality may be significant, public involvement and a formal environmental impact statement are required. As a result, some proposed actions have been denied or less-impactive alternatives have been selected because anticipated benefits did not justify the estimated environmental harm. When actions are approved, mitigating measures must be provided to reduce social and environmental impacts and ongoing projects must be monitored for compliance with these.

Clean Water Act of 1972 (amended)

Drinking water quality has improved considerably since this act was passed. The EPA set the standards and provided funds for thousands of water treatment facilities. In cooperation with local governments, EPA worked to reduce the discharge of pollution into rivers and lakes and to rehabilitate previously polluted waterways. Even though some spectacular successes have occurred, such as the partial restoration of Lake Erie, some 30 percent of rivers and 40 percent of lakes were still below standard in 1990. In April 1993, nearly 400,000 residents of Milwaukee, Wisconsin, became ill from the tap water during a change in the city's chemical treatment routine.

Providing safe water for drinking and recreation is a formidable challenge because pollution comes from so many sources, including toxic chemicals from industry, agricultural wastes, fertilizers, pesticides, bacteria and fungi, and municipal sewage (especially during floods). Individual citizens carelessly discard toxic items, such as motor oil, paint, and batteries (Ch. 5). State-of-the-art water treatment is expensive and most cities have outdated systems (4). Hence, U.S. goals for clean air and water are only partially attained and most developing nations have much more serious obstacles to overcome.

Jon Bowermaster reports that one major chemical company on the lower Mississippi River generates 3.3 million pounds of toxins each year "from acetaldehyde to xylene" that go directly into the local air, water, and earth (5). Rather than install pollution controls, he states, the company bought up the adjacent small town of Morrisville, paying each family enough to move to a safer environment. Aside from reported nausea and headaches suffered by workers and residents, communities along the river have among the highest U.S. death rates from cancer in their internal organs. The company has paid fines to the State of Louisiana and to the EPA for excessive release of toxins into the air and river but these conditions persist.

Superfund, 1980

Under the Superfund program, the EPA is required to identify abandoned toxic waste sites and to supervise their cleanup, using funds from federal and industry sources. Thousands of sites were identified and over 1200 were put on a priority list for early action, but so far only a fraction have been treated, due in part to the high cost. Millions of tons of toxic wastes are involved, many dumped illegally. Superfund's critics say the risks of doing nothing are often exaggerated and that in many cases the benefits expected from cleanup don't justify the costs.

In 1992, EPA estimated the total cost of environmental cleanup in the U.S. to be about 125 billion dollars annually, 2.4 percent of the gross national product. This estimate assumed full attainment of goals for air and water quality, solid and hazardous wastes, and toxic substances. This money, if prudently spent, goes back into the economy in the form of jobs, payments for goods and services, and increased tax revenues from resulting incomes and profits. An important nonmonetary benefit is a safer, more attractive, and healthier environment that may give our citizens a longer life, fewer days missed from work, and lower medical bills.

Endangered Species Act of 1973 (amended)

Environmentalists strongly support the Endangered Species Act, despite the adverse publicity it has received when efforts to preserve a species contribute to employee layoffs and reduced economic activity. They say these outcomes seldom occur and when they do, critics tend to blame the law and to ignore other factors involved in the slowdown. For example, efforts to protect the endangered spotted owl were blamed for the 1980s logging cutbacks in the Pacific Northwest, but other trends were also responsible. In recent years, logging and milling operations have been modernized and far fewer workers are required to produce a given amount of lumber. In addition, many people believe that the level of logging permitted on federal lands during the Reagan years was unsustainable and had to be reduced. There was also growing concern that large-scale logging, mining, related road building, and other activities were adversely affecting other wildlife populations, water quality, and commercial and recreational fishing.

The main criticism of the law is that many protected species are seen as unessential to human or to broad ecosystem needs, and preserving them may be costly for both governments and the private sector. Yet this act has undeniably saved some important species from extinction and helped many others to regain a healthy population. Examples include the bald eagle (the nation's symbol), peregrine falcon, California sea otter, grey whale, whooping crane, and brown pelican.

Of the 962 plants and animals listed as endangered or threatened in the U.S., only 11 have fully recovered, 21 are or may be extinct, 10 percent are gaining strength, and 28 percent are stable instead of declining (6). Supporters of the law say that 20 years of saving species probably had less effect on jobs and the economy than just one year (1995) of corporate downsizing to increase profits for stockholders and managers.

International Efforts

Montreal Protocol, 1987

Under this agreement, strengthened in London in 1990, countries of the world will terminate production of chlorofluorocarbons (CFCs) by 2000 to reduce depletion of the protective ozone layer in the atmosphere (Ch. 5). The U.S., the major producer, accelerated its schedule to 1995. Without further infusions of chlorine atoms into the atmosphere, ozone levels are expected to be increasing in another 30 years.

Ocean Dumping Ban, 1988

Ocean pollution is an especially serious problem because all of the world's contaminated rivers empty into the oceans. Thousands of coastal cities discharge trillions of gallons of raw sewage and wastewater directly into them, and ocean vessels sometimes spill toxic cargos (Ch. 5). Because of pollution, large areas of beaches are periodically closed to swimming and shellfish harvesting.

There are other problems as well. Nature Conservancy magazine reports that some fishermen in the western Pacific now use hundreds of tons of cyanide gas to stun and capture live fish for Asian markets. In the process, especially in the waters of Indonesia and the Philippines, they destroy both coral beds and the creatures that depend on them (7).

Efforts to reduce ocean pollution are expanding, as effluents into rivers are better controlled, more municipalities modernize their sewage treatment facilities, and more ocean dumpers are prosecuted. Even though waste dumping at sea is now prohibited, some still occurs accidentally, as with oil spills, and intentionally, as legal waste disposal becomes more difficult. When the Mediterranean Sea was said to be "dying," the 18 nations around it unanimously adopted a 1976 action plan to reduce pollution and to take other steps to enhance the environment and welfare of each country (8). Worldwide, citizens groups are getting involved in ocean cleanup, with campaigns to protect marine life and annual rallies to clear debris from the beaches. One such event, the 1991 International Coastal Cleanup, enlisted 145,000 volunteers in 12 countries to pick up 3.7 million pounds of trash, most commonly plastics, on over 4700 miles of beaches (9).

Global Warming Treaty, 1992

There is growing consensus among scientists that global warming is occurring due to carbon dioxide accumulating in the atmosphere, that human activities contribute to this, and that nations must take steps to slow this process (Ch. 5). The final report of the U.N. Intergovernmental Panel on Climate Change (IPCC) confirmed this trend in 1995, following a conference of 500 delegates from 120 nations with oil-producing countries joining in the consensus. IPCC forecasts an average global temperature increase of 3.6 degrees F. within 100 years, causing rising ocean levels, more erratic weather, and extended heat waves.

At the Rio Summit meeting in 1992, the nations of western Europe and many others pressed for a binding treaty to reduce carbon emissions to 1990 levels by the year 2000, but when President Bush refused to sign it, the agreement remained voluntary. World reaction was critical because the United States is the world's leading energy consumer and thus a major source of carbon dioxide and other atmospheric pollutants. Later, Congress and President Clinton agreed to comply with the terms of the pact, but by then some western European nations seemed less eager (10). But when delegates of these nations met in Berlin in 1995, there was again broad agreement to work toward binding ceilings on greenhouse emissions.

Global Fishing Treaty, 1995

After three years of debate, 100 nations negotiated the first treaty regulating fishing on the high seas, but it is pending until ratified by at least 30 governments. Its aim is to curb overfishing of migratory fish, such as cod, pollock, tuna, swordfish, and marlin, and to reduce the destruction of unwanted species that are caught, die in the nets, and are then discarded.

World Bank Programs

The World Bank was organized in 1945 to aid in the post-war reconstruction of Europe but its present role is to finance projects with low-interest loans to developing countries. In 1995, 176 nations were members of the bank, with the U.S., Japan, Germany, France and Great Britain providing 42 percent of its assets. Helpful as the bank may have been to many countries, it has been criticized for backing projects that were harmful to the environment and for making loans to dictatorial nations with little respect for human rights.

According to a Union of Concerned Scientists "Fact Sheet" and other sources, the World Bank has funded massive agricultural and land-clearing projects without adequate consideration of environmental effects, and some of these have required massive resettlement of residents. The World Bank says it is responding to these criticisms by more thoroughly evaluating the social and environmental impacts of their programs and by working toward sustainable development. Their critics hope this is true.

Dutch National Environmental Plan

The Netherlands is a nation with 15.5 million people on just 16,000 square miles (41,400 sq. km.) of land, almost 40 percent of it reclaimed from the ocean. Its population density is 965 per square mile, about three times that of New York State, yet it enjoys high living standards by most measures. In 1989-1990, this country adopted a long-debated National Environmental Policy Plan, an ambitious effort to integrate future development with environmental protection and to achieve a sustainable economy within 20 years.

The plan contains a vision of the future with specific goals along the way and also provides for periodic revisions. It lists 50 strategic measures intended to curb environmental deterioration, reduce pollution at the source, cut pesticide use, monitor and control wastes, conserve energy and other resources, improve product quality, upgrade mass transit facilities, and increase recycling (including composting). If this plan works reasonably well, it may serve as a model for other nations to emulate (11).

Chapter 12

Some Conclusions and a Look Ahead

This final chapter recaps and integrates major themes of this book, including the emergence of the environmental dilemma, persistent environmental issues, areas of consensus, promising responses, and indications of progress. These provide some basis for speculation about what lies in store for future generations.

The Environmental Dilemma in Review

The creation of large cohesive nations in Europe and North America over two centuries ago improved conditions for travel and trade, increased access to raw materials and markets, reduced traditional constraints on freedom of thought, and encouraged scientific discoveries. The time was ripe for the Industrial Revolution, a period of technological innovation, rapid economic growth, and remarkable social changes. Beginning in England in the early 1800s, this revolution soon diffused to other western nations and is now global in scope.

The Triumph of Technology

The Industrial Revolution and the social changes it stimulated were dramatic and far-reaching, including:

--new sources of power: coal, steam, gasoline, and electricity to provide energy for factories, heating, lighting, transportation, and communications.

--shifting the production of consumer goods from farms, homes, and shops to large urban factories that mass-produced an expanding variety of inexpensive merchandise.

--mechanizing agriculture, displacing smaller farmers and farm workers, and inducing massive rural to urban migration. --creating cities with millions of people and increasingly distinctive urban values and lifestyles.

--introducing medical procedures and social reforms that reduced the world's death rate, increased its population sixfold in 150 years, and now permit a net gain of 90 million people annually.

--inducing rising per-capita consumption of goods and services worldwide, and a rapid drain on the planet's most accessible natural resources: timber, minerals, fossil fuels, croplands, fresh water, and wildlife.

--causing environmental degradation on a global scale by expanding urban infrastructures, polluting air and water, farming marginal lands, eliminating thousands of species, and discarding enormous quantities of wastes, many of them toxic to living things.

Environmental Movement

Over a century ago, some nations began to mitigate the most obvious social and environmental problems from industrialization through laws and agency policies, but such actions were fragmentary and often weakly enforced. In the 1960s, pollution, forest and mineral depletion, erosion, species extinction, and human health effects were serious enough to kindle a global environmental movement that earned broad public support. By the mid-1990s outcomes included:

--greatly expanded research and writing on population growth, industrial development, consumption trends, and resulting physical, biological, social, and economic impacts.

--national and international efforts to curb "runaway" population growth, to reduce disease and malnutrition, and to conserve resources.

--laws and international agreements to protect and to restore the environment by monitoring conditions, mitigating pollution, reducing waste, preserving natural areas, and saving endangered species (Ch. 8).

--extensive and effective local efforts to reduce pollution, to recycle waste, and to improve the quality of lakes, streams, forests, wetlands, groundwater, and devastated sites.

Quest for Knowledge and Understanding

Environmental concerns and subsequent protective measures soon generated a countermovement and today controversy about the state of the planet and needed responses continues unabated.

Perils of Misinformation

To succeed in protecting the global environment, and hence the quality of life for future generations, we must first understand the challenges at hand and then be diligent in responding to them. When we overstate the perils that lie ahead, we risk making needless sacrifices and may waste human and financial resources that could be better used. We also lose credibility if we predict disaster and it fails to materialize. It is even more reckless to rationalize that major problems are minor until they escalate to levels of environmental deterioration and social deprivation that are wholly unacceptable and virtually irreversible. Yet each of these approaches has spokespersons and staunch defenders among businessmen, academics, technical specialists, politicians, and/or citizens at large.

At one extreme, radical environmentalists claim that the deepening global crisis requires urgent, far-reaching actions, including prompt reductions in population, conservation or preservation of remaining resources, less centralization of industry, and slower economic growth. At the opposite end, ultraconservatives deny there is a crisis, insist that most environmental regulations are unnecessary, want easy commercial access to public resources, promote continued economic growth, and oppose public support for family planning. Both factions seek to influence the great majority of Americans who have intermediate views or simply take their environment for granted.

Obstacles to Sound Science and Agreement

The most effective responses to the environmental dilemma stem from sound, unbiased, interdisciplinary assessments of social and environmental trends, their significance, and feasible changes, when needed. These studies provide a basis for greater consensus within and among nations about what must be done, who should do it, and how it should be financed.

Several barriers impede our efforts to do this, including:

--The complexity of the environmental crisis, with its numerous dimensions, and the difficulty of integrating fragmentary research from many different disciplines. Many prominent scientists perceive environmental problems mainly from the perspective of their disciplines or funding sources and do not evaluate their findings in a larger context. Many are reluctant to go beyond the limits of their expertise and to risk the scorn of experts in other fields.

--Resistance from industrial interests that equate environmental protection and conservation with reductions in profits and undervalue the future benefits of sustainable use of natural resources. Whether it is timber harvesting, wildlife protection, livestock grazing, outdoor recreation, or copper mining, some managers find it difficult or impractical to view their sector of economic activity in a larger social, environmental, and temporal context.

--Misrepresentation by commissioned researchers, think tanks, and interest groups that seek evidence to support the preordained conclusions of their supporters and scorn contradictory research results, however well-founded. Issues are also obscured by spokespersons for special interests, who freely voice "off the cuff" opinions and offer assumptions as facts.

--Strong, uncompromising reactions from people with rigid religious or ideological beliefs that enable them to justify human dominance over nature, unrestricted population growth, exploitation of natural resources, forcing "correct" views on others, and ignoring environmental consequences.

--Limited funding for environmental research and protection. Creative programs could reduce this gap by diverting both unused human resources and resulting savings to environmental programs, thus providing meaningful new jobs and additional tax revenues while ensuring a better quality of life for future generations.

Emerging Consensus

Some scientists and public officials dramatize the severity of environmental problems and advocate stern, arbitrary, and perhaps unworkable solutions. Others provide valuable feedback on emerging problems, unmet needs, spurious concerns, and ineffective actions, and help to redirect public and private efforts to maintain a quality environment. Yet a broad consensus is being forged by the majority of scientists, agency specialists, and others who are well-informed on environmental issues. Areas of broad agreement include:

1. There is a global environmental crisis due to rapid population growth, impactive technology, mounting per-capita consumption, rapid depletion

Some Conclusions and a Look Ahead 171

of many of the most accessible natural resources, and global deterioration of the natural environment. Experts often differ in their estimates of the most critical aspects and the severity of this crisis.

Thus the environmental crisis is not simply a creation of gadfly scientists with a need to "publish or perish" and others who personally benefit from promoting environmental causes, even though, in restrospect, some of them have overstated their case. Thousands of scientists have developed and now share the published findings and concerns of both the United Nations Environmental Programme and the Union of Concerned Scientists. Polls regularly conducted since Earth Day 1970 verify the environmental awareness and concern of most U.S. citizens, and recent international surveys yield similar results abroad.

A minority of the scientific community and lay citizens claim there is no environmental crisis. Among the most adamant critics are people who fear the loss of profits, jobs, funding, or prestige if overpopulation and resulting environmental problems are acknowledged and corrective actions are taken (even though some of them might eventually benefit from effective reforms).

2. Numerous authorities agree that humans have the intelligence and resources to deal with current and future environmental challenges, although some of them have concerns about the slow pace of clarifying needs and effectively responding.

3. One of the main reasons that alleged "prophets of doom," beginning with Malthus, were so pessimistic about future conditions is that they underestimated technological development. Technology has a remarkable potential for both disturbing the environment and for helping us to solve environmental problems. But our ability to forecast breakthroughs and the range of applications of technology is limited. The farther we look ahead, the more uncertain we are because social factors as well as scientific advances are involved.

This poses a dilemma for planners because, even though new inventions and discoveries may be able to resolve nagging problems for us, we cannot sit back and assume that they will occur and be constructively utilized. Yet recent developments demonstrate that expanding technology can and will be very helpful in resolving many aspects of the environmental crisis when we are willing to use it for these purposes.

4. The quality of life for future generations is speculative, but clearly the nations of the world are increasingly successful in defining and resolving mutual problems. The planet has over 200 nations with wide religious, ethnic, and social class differences within and among them. These cultural

barriers impede cooperation in identifying and making needed changes, both within the U.S. and in global programs. Established short-term goals and gratifications often receive priority, with business as usual until people tardily agree that previously avoidable dilemmas have become inevitable.

5. There is often a big difference between what a nation or the world can do and what it will do, even in response to vital needs. Social and economic factors always channel potential applications of technology and the free market is reluctant to resolve problems when profits are not foreseeable. For example, the world has the technology to produce and distribute enough food to eliminate starvation and malnutrition, but it does not. To do this, nations would need to agree on:

--who really needs and deserves food,
--which foods are needed and who can provide them, and
--who will pay for them?

Thoughtful people might also wonder:
--if recipients would become dependent on free food,
--if free food would cause their population to grow faster than their economy,
--if massive food exports would drive up food prices elsewhere,
--how food could be delivered in countries lacking adequate roads and refrigeration, and
--whether corrupt politicians, bandits, or the black market would divert food from the needy.

Rough Measures of Progress

Four current trends provide clues to the success of recent efforts to manage the global environmental crisis, namely, population growth rates, results of environmental programs, evidence of international cooperation, and new technological developments.

Population Growth Trends

Many people think rapid population growth is the most critical dimension of today's population-environment dilemma because it often intensifies other environmental conditions. Successful efforts to reduce environmental problems, such as air pollution, local famine, species extinction, juvenile crime, and drug abuse may merely reduce the rate of environmental decline if human population proliferates. But curbing and stabilizing human population growth, keeping it within global and regional carrying capacity, is a formidable task without resorting to stern measures, such as those used in China.

A century ago, large families were the rule but death rates were high and large-scale emigration to other lands was feasible. Today birth rates and death rates in most countries are lower than even a generation ago, due in part to effective measures of birth control. But because people live much longer and most babies survive to become parents, world population continues to grow at an annual rate equivalent to 14 New York Cities. In most countries, subsistence farming has been extended to its limit and growing numbers move to cities and depend on cash incomes or relief programs. Massive emigration is not feasible because the most popular destinations restrict immigration and accommodate very small fraction of the world's net population growth. Many countries also restrict employment of noncitizens and prosecute or deport illegal immigrants.

The low birth rates and small family size in industrial nations are a sharp departure from traditional beliefs and practices. In these and many other countries, family size is increasingly influenced by:
--personal career and lifestyle aspirations,
--expanded roles and more autonomy for women and children,
--the extent of formal education of prospective parents,
--access to effective birth control techniques,
--the age at which women begin to have children, and
--current economic conditions.

Success of Environmental Programs

Environmental protection and natural resource conservation programs have made significant progress during the 30 years since Rachel Carson, Fairfield Osborn, and many others alerted the world to the emerging situation. These programs are expanding in Europe, the United States, and other industrial nations, and diffusing to most of the world. Air and water quality are much improved in some areas and more materials are recycled. Many local programs are stemming or reversing decades of environmental decline at specific sites.

Yet many authorities think that these impressive gains are more than offset by additional population growth, increased consumption, and accumulating stresses on the environment. More populous future generations will have fewer resources and a more septic environment if current generations merely diminish but do not reverse destructive trends.

International Cooperation

International cooperation in programs to reduce pollution, conserve resources, regulate ocean fishing, combat hunger and epidemic diseases, and reduce the birth rate have been more forthcoming than many people thought possible a generation ago. The Cold War is over, literacy is rising, communication has improved, technology is increasingly shared globally, and many more world leaders see the need to respond to environmental concerns.

But religious, ideological, and ethnic differences are a continuing barrier to cooperation. Both dedicated leadership and clear evidence of an urgent need to act are essential to improving international cooperation.

Technological Development

Scientific and applied achievements in agriculture, food preservation, transportation, biotechnology, earth sciences, engineering, medicine, and social problem resolution have enabled far more people to survive on earth than earlier projections indicated. The worldwide electronics revolution is an important catalyst for research and applications. It permits instantaneous communication, greatly enhances our ability to assess and project trends and needs, and provides a vast capacity for data storage and retrieval. Another positive development is growing public, industry, and agency recognition of presently or potentially harmful trends that can be mitigated through changes in technology and practices.

Even though our ever-increasing technological capability will help to resolve impending environmental problems, as it has in recent decades, some limitations are also apparent. The Meadows team used complex computer models to predict future social and environmental conditions based on current trends and foreseeable changes during the same period (Ch. 9). This is a reasonable approach but even experts differ in what they foresee and the numerous options that lie ahead must be narrowed to have a workable model.

Many needed applications of technology will not occur due to limited funding. Unforeseen events are bound to occur, such as extended droughts, disastrous accidents, devastating disease epidemics, the rapid development of fusion energy, or an unexpected decline in the birth rate.

Some Perplexing Issues

The Aura of Western Abundance

Resource shortages seem inconceivable to many Americans when they see tens of thousands of items cramming the shelves and bins of a supermarket or a giant discount store, or visit an overflowing auto sales lot. To appreciate the present and future global situation, one must temper this impression of abundance with some realism. For example:

--These opulent displays exist chiefly in highly industrial countries and mostly since 1950. Many of their own citizens and the majority of the world's other people cannot afford most of what is on display.

--Earlier generations subsisted mostly on locally accessible resources, but in the last two generations industrial societies have developed and imported large quantities of raw materials and factory goods from all around the world. Producers will continue to provide these goods and services as long as they can get the needed energy and raw materials, and earn a profit. As other nations industrialize and grow in population, the drain on their resources will intensify and they may be more reluctant to export raw materials. Prices will increase if resources become scarce or more costly to produce and cheaper substitutes are not available.

The Earth's Carrying Capacity

Authorities differ widely on the number of people the earth can support adequately on a long-term, sustainable basis. Estimates range from a billion or less (world population is already five times greater) to 20 billion or more (several times the present population), reflecting the social and economic philosophy and breadth of knowledge of each forecaster.

Any realistic estimate of the planet's human carrying capacity must consider many variables, including the material standard of living desired, the current condition of the environment, the likelihood of countless future generations and their probable population, the supply of undiscovered but accessible energy and raw materials, existing and foreseeable technology, and the productive efficiency of farms, factories, forests, and fisheries. Other important factors are foreseeable social conditions and trends, the expected results of resource conservation policies, and unknowns, such as unfavorable climatic cycles, wars and embargoes, and disease epidemics. Obviously this task is highly speculative.

Currently individual nations and geographical regions vary widely in their carrying capacity for many of these reasons. Belgium and Japan, each with over 800 people per square mile but a slow rate of population growth, enjoy high material living standards by selling manufactured goods and importing what they cannot efficiently produce. Jordan and Ecuador, with only about 100 persons per square mile but more rapid growth and much less modern industry, have relatively low material living standards. With sound conservation policies, the United States could conceivably consume much less oil, metals, wood, and food and then perhaps support more people without a significant reduction in its living standard in the near future.

Limiting Population Growth

A major objective of most proposals for natural resource conservation and sustainable development is manageable population growth. Additional billions of people add to the drain on the earth's finite resources, unless major changes are initiated to accommodate more people. Even if the next two generations in rapid-growth nations retain their present level of material comfort (well below European standards in most cases), twice as many people could mean far more individual cases of crowding, malnutrition, and suffering, and a greater assault on the natural environment. People in all countries of the world want more than they now have and, barring major calamities, hundreds of future generations will too.

China's Coercive Population Policy

Opponents of U.S. support for family planning programs point out that the rate of world population growth is now slackening. Many of these same people are extremely critical of China's use of coercive measures to ensure slower growth. Ironically, China's harsh program is the main reason for the reduction of total world growth rates from 2.0 to 1.7 percent annually, since it has a fifth of the world's people.

Some U.S. politicians favor including China's family size restriction as a basis for asylum. This move could encourage millions of rural Chinese to emigrate to the U.S., enable them to have more children, and increase their per-capita consumption.

The Plight of Least-Developed Nations

Extremely rapid population growth during the past few decades has altered the distribution of the world's people and changed the political and

Some Conclusions and a Look Ahead   177

economic relationships of nations. It has disrupted many of the planet's ecological systems and may be delaying or precluding higher living standards in many developing nations.

The outlook is least favorable in countries that are densely populated in relation to their arable land and other resources, continue to double in population each generation, and have limited industrial development. Rapid population growth places severe strains on these societies. Many of them are politically unstable, have high rates of illiteracy, are beset by internal conflicts, and lack the means (such as oil revenues) to industrialize or to import food and other necessities from abroad. When growth is primarily due to births, half of the population may be under 16 years of age. These children need food, clothing, schooling, medical care, and other essentials, but they outnumber employed parents and other taxpayers who must provide for them. As they mature, their continuing needs for cropland, jobs, housing, schools, public services, and utilities may well exceed the resources and productive capacity of their country.

Estimates vary, but at least a fifth of the world's people, chiefly in less-developed countries, are jobless, malnourished, and/or lack adequate housing and medical care. Their plight is now more visible in the West due to the vivid images of environmental degradation, massive dislocation, and starvation presented by the media and observed by overseas travelers. Many of them are sites of mines and factories developed and owned by western industrial nations seeking cheaper labor and fewer restraints on harmful environmental practices. Most of these countries want more industries and are exporting their minerals, timber, and even foodstuffs at bargain basement prices to obtain the capital for investment.

Costs of Environmental Reform

Although environmental reforms sometimes cost jobs and reduce profits, they also create new businesses and occupations. Reform advocates claim that most programs require fairly minor adjustments in the total pattern of business activity. More importantly, needed reforms help to ensure that the next and succeeding generations will have a healthful environment and more abundant supplies of natural resources.

Economic hardships do occur, as when many sawmills and logging operations in the western U.S. were closed due to reductions in National Forest timber available for harvest, ostensibly to save the northern spotted owl, an endangered species. The companies most affected were small with few or no private timber reserves. After decades of heavy logging on both federal and private lands, and the substitution of machines for workers, many closures had already occurred or were impending but the owl took the brunt of the blame.

Conservationists say that continuing logging (ignoring the owl) might have delayed some closures for from several months to a few years but only at the cost of losing the remaining stands of old-growth forest, increased erosion and flooding, water pollution, fishery depletion, and other ecosystem disruption.

Costs of a Stable Population

Some economists claim that family planning to stabilize population levels may generate serious economic problems, such as labor shortages, reduced industrial output, the phasedown of youth-oriented businesses and institutions, and inadequate revenues to care for the elderly. They seem to imply that population growth must continue forever to ensure a healthy economy. But computers, robots, and other sophisticated machines are replacing workers and many young people now have trouble finding jobs that offer a living wage and a promising career. Most countries today with little or no population growth seem to have an adequate labor force and enjoy material living standards well above the world average. Compared to fast-growth nations, they have fewer mouths to feed in each family, a greater percentage of taxpaying citizens, and less need for crash programs to build roads, schools, hospitals, and housing to accommodate additional people. They have the financial resources needed to improve their environment and quality of life.

As for financing retirement and Medicare, new conditions require innovative adaptations. Many workers and their employers could provide for their future through increased lifetime savings, such as individual (IRA) and Keogh retirement plans.

The Need for American Leadership

Due to innovations in transportation, communication, and trade, the world has become "smaller" and more interdependent. The actions of one nation increasingly affect others. Most countries are intimately involved with the rest of the world through mutual defense pacts, international programs, and family members working abroad. The U.S. cannot isolate itself from the world's problems, but can benefit from a leadership role in analyzing and resolving them. It is the world's richest nation, has the best research facilities, is heavily involved in international commerce, and is also the leading per-capita consumer and polluter.

Americans must be willing to plan ahead for the needs of future generations, to take corrective actions when necessary, and to cooperate in international efforts to protect the environment, to reduce poverty and disease, and to work toward sustainable development. To do less is to jeopardize the health and well-being of the globe, including the U.S.

And What Lies Ahead?

Scientists, futurists, visionaries, fiction writers, and more firmly-anchored professional planners have developed hundreds of scenarios of what the nation and world will be like in 50, 100, or 1000 years. Some of their expectations are fanciful, others are feasible, and many of them are uncertain for reasons expressed earlier.

The Fanciful

Some futurists envision artificial islands, mile-high skyscrapers, and space stations to house people. Perhaps some day we will have some of these, because we already have most of the technology to do it. But for the foreseeable future the cost of construction is so great that, unless they were subsidized by the taxpayers, only millionaires could afford to live on them and they may prefer to live elsewhere.

Other people suggest reducing world population pressures by increasing migration to less-crowded countries. Even if relatively uncrowded United States, Canada, Russia, and Australia were willing to admit 57 million more immigrants (much less than the increase each year), only one percent of the world's people would thus be accommodated and conditions in their home countries would not change much. But for the nations receiving this number (equal to the entire population of Great Britain or France), the resulting problems of cultural assimilation, the stress on schools, public services, housing, utilities, roads, and other facilities, plus tax increases and inflation of land and housing values could be formidable.

Interplanetary migration, proposed by some visionaries, is not now a viable option and may never be a technically feasible means of transporting hundreds of millions of people. Certainly we cannot count on it. If we learned how to make Mars habitable, it could take centuries to complete the job and at present growth rates, we could populate it in one generation.

If we found a habitable planet in another solar system, we would need the materials, energy sources, and funding to produce and propel thousands of enormous spacecraft at something approaching the speed of light (186,000 miles a second). Then we would need millions of people willing to leave this planet, to travel for many earth-years, and to learn to cope with an exceedingly different environment.

The Feasible

A better course is to keep population growth in check and to maintain environmental quality. Visitors who have witnessed the environmental devastation in eastern Europe and the former Soviet Union may doubt that a coercive one-party system is the solution. Others who have read of the environmental damage and waste that occurred during the "robber baron" era a century ago or have watched seven tobacco company executives under solemn oath telling a congressional committee that cigarettes are not addictive may conclude that an unregulated free market is also not the answer.

Making our peace with the environment requires cooperative efforts and innovative approaches, large and small:

--building products that last, avoiding excessive packaging, and promoting standardization and reuse of parts and containers.

--adjusting reward structures (tax incentives, bonuses, recognition) to encourage conservation and prudence rather than waste, excess, and short-term gains at the expense of the future.

--using taxes to make pollution and waste expensive; e.g., energy or emissions taxes.

--involving everyone, especially young people and the unemployed in meaningful activities that protect and restore the environment and promote responsible parenthood.

--improving and extending programs to reduce famine and epidemic diseases.

--discouraging conception of unwanted, potentially neglected and deprived children.

--assisting nations to become self-sufficient and to protect their environment, which will in turn improve both their quality of life and the global environment.

--increasing the authority, functions, and autonomy of the Council on Environmental Quality or an equivalent body, so it can assess conditions and trends, formulate and revise national goals, monitor agency compliance with environmental laws, coordinate involvement in international efforts, and recommend needed changes to the President and Congress.

Appendix A: Population Data

Table A-1

The World's Largest Metropolitan Cities

City, Country	Millions	City, Country	Millions
Tokyo, Japan	26.8	Osaka, Japan	10.6
Sao Paulo, Brazil*	16.4	Lagos, Nigeria**	10.3
New York, USA	16.3	Rio de Janeiro, Brazil	9.9
Mexico City	15.6	Delhi, India*	9.9
Bombay, India**	15.1	Karachi, Pakistan**	9.9
Shanghai, China*	15.1	Cairo, Egypt*	9.7
Los Angeles, USA	12.4	Paris, France	9.5
Beijing, China*	12.4	Manila, Philippines*	9.3
Calcutta, India	11.7	Moscow, Russia	9.2
Seoul, Korea	11.6	Dhaka, Bangladesh**	7.8
Jakarta, Indonesia**	11.5	Istanbul, Turkey*	7.8
Buenos Aires, Argentina	11.0	Lima, Peru*	7.5
Tianjin, China*	10.7		

Annual growth rate exceeds: * 2.0 percent
                            ** 4.0 percent

Source: United Nations Population Divison, 1994 data, as quoted in *World Resources 1996-1997*, **Table l.l.**

Table A-2

Population Trends in Selected Countries of the World

	Population in millions			Annual %	Density
	1950	1995	2025	increase	sq.mi.
Location					
World	2520	5716	8294	1.6	113
North America	220	454	616	1.4	48
Canada	14	29	38	1.2	8
Costa Rica	1	3	6	2.4	175
Cuba	6	11	13	0.8	256
Guatemala	3	11	22	2.9	263
Haiti	3	7	13	2.0	613
Mexico	28	94	137	2.1	127
Nicaragua	1	4	9	3.7	91
United States	152	263	331	1.0	75
South America	112	320	463	1.7	47
Argentina	17	35	46	1.2	32
Brazil	53	148	230	1.7	49
Chile	6	14	20	1.6	49
Colombia	12	35	49	1.7	90
Peru	8	24	37	1.9	49
Venezuela	5	22	35	2.3	62
Europe	549	727	718	0.2	190
France	42	58	61	0.4	276
Germany	68	82	76	-0.2	601
Hungary	9	10	9	-0.5	289
Italy	47	57	52	0.1	513
Netherlands	10	15	16	0.7	1179
Poland	25	38	42	0.1	330
Russia	103	147	139	-0.1	23

Appendix A

Location	Population in millions			Annual % increase	Density sq.mi.
	1950	1995	2025		
Spain	28	40	38	-0.1	204
Sweden	7	9	10	0.5	56
Ukraine	37	51	49	-0.1	223
United Kingdom	51	58	61	0.3	625
Africa	224	728	1496	2.8	62
Algeria	9	28	45	2.3	31
Angola	4	11	27	3.7	21
Egypt	22	63	97	2.2	162
Ethiopia	18	55	127	3.0	127
Kenya	6	28	63	3.6	131
Libya	1	5	13	3.5	8
Madagascar	4	15	34	3.2	62
Mozambique	6	14	35	2.4	60
Nigeria	33	112	238	3.0	288
South Africa	13	41	71	2.2	96
Sudan	9	28	58	2.7	33
Tanzania	8	30	63	3.0	84
Uganda	5	21	48	3.4	254
Zaire	12	44	105	3.2	50
Asia	1403	3458	4960	1.6	187
Afghanistan	9	20	45	5.8	85
Bangladesh	42	120	196	2.2	2478
China	555	1221	1526	1.1	334
India	358	936	1392	1.9	816
Indonesia	80	198	276	1.6	289
Iran	16	67	124	2.7	102
Iraq	5	20	43	2.5	123
Israel	1	6	8	2.5*	655
Japan	84	125	122	0.3	823
Korea (ROK)	20	45	54	1.0	1202
Malaysia	6	20	32	2.4	155
Myanmar	17	47	76	2.1	178

	Population in millions			Annual %	Density
	1950	1995	2025	increase	sq.mi.

Location					
Pakistan	39	140	285	2.8	438
Philippines	21	68	105	2.1	636
Syria	3	15	34	3.4	217
Thailand	20	59	74	1.1	305
Turkey	21	62	91	2.0	213
Viet Nam	30	75	118	2.2	592
Oceania	13	29	41	1.5	9
Australia	8	18	25	1.5	6
New Zealand	2	4	4	1.2	33
New Guinea	2	4	8	2.3	25

Sources: *World Resources 1996-1997.* Table 8-1
Statistical Abstract of the United States, 1995.
Table 1361.   *Inconsistent data

Appendix B

International Environmental Initiatives

A few noteworthy international agreements to protect the environment occurred earlier this century, including limitations on seal harvesting in the north Pacific (1911), the International Whaling Convention to reduce the slaughter of whales (1946), and the Antarctica Treaty to protect that continent from undue exploitation (1959).

Since 1972, international cooperation to protect the environment has steadily increased. Each decade more nations are involved in resolving a growing number of issues. The following is a selection of important events held and actions taken.

United Nations Conference on the Human Environment, 1972

Representatives of 113 nations met to develop the United Nations Environment Programme for international action to protect the global environment.

London Dumping Convention, 1972

Ultimately 66 nations outlawed dumping radioactive and other dangerous wastes into the oceans. Amended to ban other industrial wastes by 1995.

Convention on International Trade in Endangered Species, 1973

Protects endangered species by restricting the import and export of endangered plants and animals and products made from them.

Habitat I, 1976

Held in Vancouver, Canada. Addressed variety of social issues, including the need for adequate shelter, human dignity, social justice, and freedom of movement.

Commercial Whaling Ban, 1982

Nations agree to ban all commercial whaling.

Montreal Protocol, 1987

Agreement by 24 nations to reduce chlorofluorocarbon (CFC) production by 50 percent by 1999 (to diminish depletion of the earth's protective ozone layer); later amended to stop producing CFCs by 1999.

Antarctica Treaty, 1991

Amends the 1959 Antarctica agreement by banning all mineral exploration and development for 50 years, protecting wildlife, and regulating waste disposal and pollution.

United Nations Conference on Environment and Development, 1992

The 1992 Earth Summit in Rio de Janeiro, Brazil, was a crucial step in international efforts to acknowledge and to ameliorate the emerging environmental dilemma. Nearly 100 heads of state and thousands of environmental scientists and practitioners met to plan the reduction of the Earth's environmental deterioration. Most participants agreed that impacts in one part of the world increasingly affect the rest of the planet and that sustainable development practices are essential if the earth is to continue to support life as we know it.

The conference reviewed the range of environmental problems of the world and adopted Agenda 21, a comprehensive statement of what needs to be done. In the words of Senator Paul Simon, it is a global plan to confront and overcome the ecological problems of the Twentieth Century. He said, "what is at stake is nothing less than the global survival of humankind."

The Preamble of Agenda 21 reads as follows: "Humanity stands at a defining moment in history. We are confronted with a perpetuation of disparities between and within nations, a worsening of poverty, hunger, ill health, and illiteracy, and the continuing deterioration of the ecosystems on which we depend for our well-being. However, integration of environment and developing concerns and greater attention to them will lead to the fulfillment of basic needs, improved living standards for all, better protected and managed ecosystems, and a safer, more prosperous future. No nation can achieve this on its own; but together we can--in a global partnership for sustainable development."

Agenda 21 is a world action plan based on the assumption that sustainable development is essential to the survival of humanity. It foresees a global civilization that exists in harmony with individual nations and prescribes measures for attaining this. It proposes efficient use of resources, effective management of pollution and wastes, reduced population growth, and acceptable living standards for all of the world's nations. It devotes 40 sections to concerns and sets forth 120 action programs to achieve its goals. The plan acknowledges that these are difficult tasks requiring major changes, new priorities, a large commitment of resources, and the unstinting cooperation of all of the world's nations and their institutions. But it is also an urgent task that must be accomplished, beginning immediately.

Agenda 21 is organized around seven major themes:

1. Quality of life. Improve the quality of life in poverty-stricken regions, reduce wasteful consumption in other areas, and stabilize population growth, consistent with the Earth's capacity to support life on a sustained basis.

2. Efficient use of resources. Manage natural resources to ensure better conservation and more efficient use of land, seas, fresh water, biological and genetic resources, and energy. Protect deserts, mountains, forests, and fragile ecosystems and provide adequate food supplies through sustainable agriculture. Reduce dependence on fossil fuels because they are limited in supply and are a major source of pollution.

3. Protection of the "global commons." Protect the atmosphere and oceans to ensure breathable air, a stable climate, preservation of the ozone layer, and abundant and edible seafood. 4. Management of human communities. Almost half of the world's people now live in urban areas, many of which are mushrooming in size, and they need both cash incomes and adequate public services to survive.

5. Chemicals and waste management. Chemicals, factory goods, and packaging materials have become essential to modern life, but major reforms are needed in their production and use. If present levels of pollution and waste in industrial nations were extended to developing nations, there would be intolerable consequences for humans and plant and animal life.

6. Sustainable economic growth. Make sustainable development the guiding principle in industry and agriculture. We must adapt accounting procedures to include the true costs of production including pollution, soil and groundwater contamination, erosion, health effects, etc.

7. Provisions for plan implementation. To succeed, the action plan requires effective coordination and the willing participation of governments at all levels, other organizations, and individual citizens. Environmental quality and sustainability must become key considerations in all decisions involving the resources of the planet.

Convention on Biological Diversity, 1992

Establishes a framework for conserving biological diversity, sustainable use of species, and sharing of genetic resources (118 parties).

Convention on Climate Change, 1992

Industrial countries are to stabilize carbon dioxide levels at 1990 levels by 2000; developing nations are to do emissions inventories; all are to work toward eventual stabilization (136 parties).

Human Rights Conference, 1993

Held in Vienna, Austria.

Convention on Desertification, 1994

Implements strategies for sustainable management of land and water, encourages national action programs, and provides guidance for local projects (106 signers, 1 party).

United Nations Conference on Population and Development, 1994

Held in Cairo, Egypt, this meeting is the third in a series of U.N. conferences dealing with population issues, following the Mexico City conference in 1983 and the Bucharest 1974 conference. Dealt with the urgency of empowering women and reducing gender bias, basing population programs on human needs rather than on impersonal target populations, family planning programs, integrating population concerns into all development policies and programs, and the extent of resources needed to provide reproductive health program worldwide.

World Conference on Women, 1995.

About 5000 delegates from 189 countries attended this fourth women's rights conference, held in Beijing, China. The objectives were to review what has been done earlier and to adopt a Platform for Action focusing on obstacles to the advancement of the majority of women in the world, and to include planks on awareness-raising, literacy, poverty, health, violence, refugees, technology, and program implementation procedures.

The final declaration of the conference urges governments of the world to observe the following principles:

--women should have freedom of choice on sexual participation and childbearing, and not suffer discrimination because they are mothers,
--genital mutilation, physical abuse, and sexual harassment are violations of human rights,
--wartime rape of women is a crime and perpetrators are war criminals,
--the family is the basic unit of society and should be strengthened, protected, and supported,
--women everywhere endure unfair discrimination and deserve access to credit, banking services, and equal rights to inherit.

Social Development Conference, 1995

Held in Copenhagen, Denmark.

Habitat II: United Nations Conference on Human Settlements, 1996

Held June 1996 in Istanbul, Turkey. Focused on the urban environment, the health and quality of life of urban residents, and the impacts of urban areas on the surrounding environment. Sought to go beyond conceptual issues, such as the right to adequate housing, to practical action that will help nations to achieve this goal.

Major sources:

Columbia Encyclopedia, 1993.
French, Hilary F. 1995. *Partnership for the Planet: An Environmental Agenda for the United Nations.* Worldwatch Paper 126. Washington, DC: Worldwatch Institute.
Sitarz, Daniel, Ed. 1993. *Agenda 21: The Earth Summit Strategy to Save Our Planet.* Boulder, CO: Earth Press.
World Resource Institute. 1996. *World Resources: A Guide to the Global Environment.* New York: Oxford University Press.
Various press reports.

Endnotes and Pertinent Readings

Introduction and Overview

Endnotes:
1. The World Resources Institute, et.al. 1996. *World Resources: A Guide to the Global Environment.* New York: Oxford University Press. Table 8.1.

Pertinent Readings:
Attenborough, David. 1984. *The Living Planet: A Portrait of the Earth.* Boston: Little, Brown and Co.
Brown, Lester R. 1984-to date. *State of the World.* An annual Worldwatch Institute report on a wide variety of population and environmental issues. Supplemented since 1992 by *Vital Signs: The Trends that Are Shaping Our Future.* New York: W.W. Norton & Co.
Commoner, Barry. 1990. *Making Peace with the Planet.* New York: Pantheon Books.
Esterbrook, Greg. 1995. *A Moment on Earth: The Coming Age of Environmental Optimism.* New York: Viking-Penguin.
Global Tomorrow Coalition. 1990. *The Global Ecology Handbook: What You Can Do About the Environmental Crisis.* Boston: Beacon Press.
Gore, Al. 1993. *Earth in the Balance: Ecology and the Human Spirit.* New York: Plume-Penguin.
Langone, John. 1992. *Our Endangered Earth: What we Can Do to Save It.* Boston: Little, Brown & Co.
Seager, Joni. 1990. *State of the Earth Atlas.* New York: Simon and Schuster.

Chapter 1: Revolutions in Technology

Endnotes:
1. Through photosynthesis, green plants convert sunlight into chemical energy which is used to transform carbon doxide and water into sugar and enables plants to

grow. Insects eat the plants, lizards eat the insects, and eagles eat the lizards. This is a food chain. Alternatively, grazing animals eat the plants and predators (including humans) eat them. Eagles, lions, humans, and others at the top of the food chain benefit from the lower levels but may also accumulate toxic substances ingested by plants and animals below them on the food chain.

2. The inspiration for this analysis of economic systems stems from the work of social anthropologist Paul Bohannon.

3. Hammond, J.L. and Barbara Hammond. 1966. *The Rise of Modern Industry*. London: Methuen & Co.

4. *Statistical Abstract of the United States*. 1995. Table 44.

5. Many developing nations were once called underdeveloped nations, implying that their development was insufficient. Third World was another term applied to nations that were not in the First World (industrial democracies) or the Second World (the communist bloc). Hence, most developing nations were also Third World nations. Since the demise of the communist bloc these terms seem to be declining in use.

6. *Statistical Abstract of the United States*. 1995. Table 1335.

7. *World Almanac, 1996*.

Pertinent Readings:

Bunch, Bryan, and Alexander Hellemans, Eds. 1993. *The Timetables of Technology: A Chronology of the Most Important People and Events in the History of Technology*. New York: Simon and Schuster.

Commoner, Barry. 1990. *Making Peace with the Planet*. New York: Random House-Pantheon Books.

Goudie, Andrew. 1994. *The Human Impact on the Natural Environment*, 4th Ed. Cambridge, MA: MIT Press.

Sclove, Richard E. 1995. *Democracy and Technology*. New York: Guilford Press.

Zey, Michael G. 1994. *Seizing the Future*. New York: Simon and Schuster.

Chapter 2: Rapid Population Growth

Endnotes:

1. Population Institute. 1995. *World Population Overview*. Reviewed in the Salem, OR, Statesman Journal, December 28, 1995.

2. Malthus, Thomas R. 1798, revised 1803. *An Essay on the Principles of Population*. London: J. Hohnson.

3. *Statistical Abstract of the United States*. 1995. Table 1361.

4. World Resources Institute. 1990. *World Resources: A Guide to the Global Environment, 1990-1991*. New York: Oxford University Press. Pp. 49-50, 254.

5. Wattenberg, Ben J. 1987. The Birth Dearth. New York: Pharos Books.

Pertinent Readings:
Brown, Lester R. and Hal Kane. 1994. *Full House: Reassessing the Earth's Population Carrying Capacity.* New York: W.W. Norton & Co.
Ehrlich, Paul R. and Anne H. Ehrlich. 1990. *The Population Explosion.* New York: Simon & Schuster.

Chapter 3: Escalating Consumption of Resources

Endnotes:
1. World Resources Institute. 1996. *World Resources: A Guide to the Global Environment, 1996-1997.* New York: Oxford University Press. Table 7.1., 7.2, and pp. 161-162. Also World Almanac, 1996. Pp. 737-837.
2. World Resources Institute. 1990. *World Resources: A Guide to the Global Environmment, 1991-1992.* New York: Oxford University Press.
3. *Statistical Abstract of the United States.* 1995. Tables 1380-1381.
4. Durning, Alan T. 1993. "Long on Things," in *Sierra* 78:1 and *Statistical Abstract of the United States,* 1991 and 1995. Tables 1399-1400.
5. Weaver, Kenneth F. 1981. "Our Energy Predicament," in *National Geographic: Special Report* (Feb 1981).
6. *Statistical Abstract of the United States.* 1995. Table 1337.
7. Worldwatch Institute estimates based on the *United Nations Statistical Yearbook* and other sources.
8. *Statistical Abstract of the United States.* 1995. Table 706.
9. Stutzman, Crispin. 1994. "Automobiles, Consumption and Sustainability," in *Re:Action*, a newsletter of the Union of Concerned Scientists (Winter 1994). Also *Statistical Abstract of the United States.* 1995. Table 1023.
10. *Statistical Abstract of the United States.* 1995. Tables 369-370.
11. *Ibid.* Tables 1378-1380.
12. *Ibid.* 1995. Table 723.
13. Tunali, Odil. 1996. "A Billion Cars: The Road Ahead." *World Watch Magazine* 9:1 (Jan-Feb 96) Pp. 24-33.
14. Durning, Alan. 1992. *How Much Is Enough: The Consumer Society and the Future of the Earth.* Worldwatch Environmental Alert Series. New York: W.W. Norton and Co.
15. *World Almanac* 1996. P. 263.
16. Motavalli, Jim. 1996. "Enough!" *E: the Environmental Magazine.* 7(2): 28-31.

Pertinent Readings:
Brown, Lester R. and Christopher Flavin. 1996. "China's Challenge to the United States and to the Earth." *World Watch* 9(5): 10-13.
Harrison, Paul. 1993. "Sex and the Single Planet." *The Amicus Journal.* 15(4): 16-26.
Shorris, Earl. 1994. *A Nation of Salesmen: The Tyranny of the Market and*

Subversion of Culture. New York: W.W. Norton & Co.

Young John E. 1991. *Discarding the Throwaway Society.* Worldwatch Paper 101. Washington, DC: Worldwatch Institute.

Young, John and Aaron Sachs. 1995. "Creating a Sustainable Materials Economy," in Worldwatch Institutes's *State of the World, 1995.* New York: W.W. Norton and Co.

Chapter 4: Depletion of Natural Resources

Endnotes

1. Boulding, Kenneth E. 1968. "The Great Society in a Small World--Dampening Reflections from the Dismal Science." In Bertram M. Gross, Ed., *A Great Society?* New York: Basic Books. P. 219.

2. Young, John E. and Aaron Sachs. 1995. "Creating a Sustainable Materials Economy," in Worldwatch Institute's *State of the World.* New York: W.W. Norton and Co.

3. Young, John E. *Mining the Earth.* 1992. Worldwatch Paper 109. Washington, DC: Worldwatch Institute. Pp. 7-8.

4. *Ibid.*, P. 9. Also: *Statistical Abstract of the United States.* 1995. Tables 1397, 1398.

5. Young, John E. 1992. Cited above. P. 29.

6. *Statistical Abstract of the United States.* 1995. Tables 1189, 1195.

7. *Ibid.* Table 1197.

8. *Ibid.* Tables 935, 936. Also: World Resources Institute. 1996. *World Resources: A Guide to the Global Environment, 1966-1967.* New York: Oxford Press. Pp. 275-27.

9. *Statistical Abstract of the United States.* 1995. Table 957.

10. World Resources Institute, cited above. 1996. P. 201.

11. *Statistical Abstract of the United States.* 1995. Tables 1152-1156. 1161-1162.

12. World Resources Institute, 1996. Cited above. P. 205.

13. Goudie, Andrew. 1994. The Human Impact on the Natural Environment. Cambridge, MA: MIT Press. Pp. 110-111.

14. Postel, Sandra. 1900. "Saving Water for Agriculture." *State of the World, 1990.* New York: W.W. Norton. Pp. 40-41.

15. *Statistical Abstract of the United States.* 1995. Table 369.

16. Russell, Dick. 1995. "Fishing Down the Food Chain." *The Amicus Journal.* 17(3): 16-24.

17. *Statistical Abstract of the United States.* 1995. Table 1163, 1165.

18. *Ibid.* Table 1168.

19. *Ibid.* Table 399.

20. *Ibid.* Table 398, 402.

Pertinent Readings:
Abramovitz, Janet N. 1996. *Imperiled Waters, Impoverished Future: The Decline of Freshwater Ecosystems*. Worldwatch Paper 128. Washington, DC: Worldwatch Institute.
Ciemins, Thomas. 1996. "Parks in Peril." *E: The Environmental Magazine*. 7(2): 36-41.
Gardner, Gary. 1996. *Shrinking Fields: Cropland Loss in a World of Eight Billion*. Worldwatch Paper 131. Washington, DC: Worldwatch Institute.
Postel, Sandra. 1992. *Last Oasis: Facing Food Scarcity*. Worldwatch Environmental Alert Series. New York: W.W. Norton & Co.
Russell, Dick 1996. "Vacuuming the Seas." *E: The Environmental Magazine*. 7(4): 28-35.
Weber, Peter T. 1994. *Net Loss: Fish, Jobs, and the Marine Environment*. Worldwatch Paper 120. Washington, DC: Worldwatch Institute.
World Resources Institute. 1996. *World Resources, 1996-1997*. New York: Oxford University Press.

Chapter 5: Environmental Degradation

Endnotes:
1. Part of the preceding discussion is based on: Wenner, Lambert N. 1992. *Minerals, People, and Dollars: Social, Economic, and Technological Aspects of Mineral Resource Development*. Missoula, MT: USDA, Forest Service, Northern Region.
2. Young, John E. 1992. *Mining the Earth*. Worldwatch Paper 109. Washington, DC: Worldwatch Institute.
3. United Nations Food and Agriculture Organization (FAO). 1995. *Forest Resources Assessment 1990: Global Synthesis*. Rome: FAO. Pp. 12-17.
4. Goudie, Andrew. 1994. *The Human Impact on the Natural Environment*. Cambridge, MA: MIT Press. Pp. 45-47.
5. Flavin, Christopher and Odil Tunali. 1996. *Climate of Hope: New Strategies for Stabilizing the World's Atmosphere*. Worldwatch Paper 130. Washington, DC: Worldwatch Institute.
6. Thompson, Jon. 1991. "East Europe's Dark Dawn." *National Geographic*. 179(6): 36-69.
7. World Resources Institute. 1996. *World Resources: A Guide to the Global Environment 1996-1997*. P. 316.
8. Goetze, Darren. 1996. "The Climes are A-Changing." *Nucleus*. 18(1): 2-3.
9. Flavin, Christopher. 1990. "Slowing Global Warming." *State of the World, 1990*. Washington, DC: Worldwatch Institute. Pp. 18-19.
10. Sampat, Payal. 1996. "The River Ganges' Long Decline." *World Watch*. 9(4): 25-32.
11. United Nations Environment Programme. 1995. *Global Biodiversity Assessment*. Cambridge, U.K.: Cambridge University Press. P. 118. Reported by World Resources Institute. 1996. Cited above. Pp. 247-252.

12. Wilson, Edward. 1993. *The Diversity of Life*. Cambridge, MA: MIT Press. More recent data were reported in the Portland Oregonian. November 14, 1995.
13. Jones, Robert F. 1990. "Farewell To Africa." *Audubon*. Sept: 50-105.
14. Harrison, George H. 1992. "Is There a Killer in Your House?" *National Wildlife* 30(6): 10-13.
15. Young, John E. 1991. *"Discarding the Throwaway Society,"* Worldwatch Paper 101, p. 16. Washington, DC: Worldwatch Institute.
16. Lenssen, Nicholas. *"Nuclear Waste: The Problem That Won't Go Away."* 1991. Worldwatch Paper 106. Washington, DC: Worldwatch Institute.

Pertinent Readings

Asimov, Isaac, and Frederik Pohl. *Our Angry Earth*. New York: Tom Doherty Associates.

Broad, Robin, and John Cavanaugh. 1994. "Borneo on the Brink: Of Rainforests and Robber Barons." *The Amicus Journal*. 16(2): 18-25.

Carrier, Jim, and Jim Richardson. "The Colorado: A River Drained Dry." *National Geographic*. 179(6): 2-35.

Devine, Robert. 1992. "The Salmon Dammed." *Audubon*. Sept 1990: 50-105. Also: 1994. "Botannical Barbarians." *Sierra*. 79(1): 50-57.

Durning, Alan T. 1993. *Saving the Forests: What Will It Take?* Worldwatch Paper 116. Washington, DC: Worldwatch Institute.

Duvall, Bill, Ed. 1994. *Clearcut: The Tragedy of Industrial Forestry*. San Francisco: Sierra Club/Earth Island Press.

Easterbrook, Greg. 1995. *A Moment on the Earth: The Coming Age of Environmental Optimism*. New York: Penguin (Viking) Books.

Ehrlich, Paul R. and Anne H. Ehrlich. 1990. *The Population Explosion*. New York: Simon & Schuster.

Gore, Al. 1993. *Earth in the Balance: Ecology and the Human Spirit*. New York: Penguin (Plume) Books.

Gribbin, John. 1990. *Hothouse Earth: The Greenhouse Effect and Gaia*. New York: Grove Weidenfeld.

Hardy, John T. 1991. "Where the Sky Meets the Sea." *Natural History*. May 91: 59-65.

Langone, John. 1992. *Our Endangered Earth*. Boston: Little, Brown and Co.

Lawren, Bill. 1992. "Singing the Blues for Songbirds." *National Wildlife*. 30(5): 4-11.

Line, Les. 1996. "Lethal Migration." *Audubon* 98(5): 50-56. McGibben, Bill. 1996. "What Good is a Forest?" *Audubon* 98(3): 34-53.

Morganstern, Richard D. and Dennis Tirpak. 1990. "The Greenhouse Gases." *EPA Journal* 16(2): 8-10.

Nature Conservancy. 1994. Special magazine issue:"Understanding Biodiversity." 44(1).

Platt, Anne E. 1995. "Dying Seas." *World Watch*. 7(1): 10-19.

Revkin, Andrew. 1992. *Global Warming: Understanding the Forecast*. New York: Abbeville Press.

Roush, G. Jon. 1989. "The Disintegrating Web: The Causes and Consequences of Extinction." *Nature Conservancy Magazine*. 39(6): 4-15.

Russell, Dick. 1996. "Vacuuming the Seas." *E: The Environmental Magazine* 7(4): 28-35.

Ryan, John C. 1992. *Life Support: Conserving Biological Diversity*. Worldwatch Paper 108. Washington, DC: Worldwatch Institute.

Schneider, Stephan. 1989. *Global Warming*. San Francisco: Sierra Club Books.

Seager, Joni. 1990. *The State of the Earth Atlas*. New York: Touchstone/Simon and Schuster.

Time. 1995. Cover Story: "The Rape of Siberia." 146(10). Also: *Time*. 1991. Special Issue "California: The Endangered Dream." 138(20).

Weber, Peter. 1993. *Abandoned Seas: Reversing the Decline of the Oceans*. Worldwatch Paper 116. Washington, DC: Worldwatch Institute.

Weiner, Jonathan. 1990. *The Next One Hundred Years: Shaping the Fate of Our Living Earth*. New York: Bantam Books. Also: 1986. *Planet Earth*. New York: Bantam Books.

Chapter 6: Social Effects of Industrial Development

Endnotes:

1. Young, John E. and Aaron Sachs. 1994. *The Next Efficiency Revolution: Creating a Sustainable Materials Economy*. Worldwatch Paper No. 121. Washington, DC: Worldwatch Institute, Pp. 20-26.
2. *Statistical Abstract of the United States*. 1995. Table 308.
3. *Ibid.* Table 349.
4. *Ibid.* Tables 6,9.
5. *Ibid.* Tables 6-11.
6. *Ibid.*, Tables 94, 100.
7. Sachs, Aaron. 1995. *Eco-Justice: Linking Human Rights and the Environment*. Worldwatch Paper 127. Washington, DC: Worldwatch Institute.

Pertinent Readings:

Buzzworm. 1993. Special Ecotravel Issue 5(2).

Durning, Alan. 1993. "Supporting Indigenous Peoples." *State of the World*. New York: W.W. Norton & Co. Pp. 80-100.

Jacobson, Jodi L. 1993. "Closing the Gender Gap in Development." *State of the World*. New York: W.W. Norton & Co. Pp. 61-79.

Kane, Hal. 1995. *The Hour of Departure: Forces that Create Refugees and Migrants*. Worldwatch Paper 125. Washington, DC: Worldwatch Institute.

Le Master, Dennis C. and John H. Beuter, Eds. 1989. *Community Stability in Forest-Based Economies*. Portland: Timber Press.

Moore, Curtis. 1995. "Green Revolution." *Sierra*. 80(1): 50.

Newland, Kathleen. 1994. "Refugees: The Rising Flood." *World Watch Magazine* 7(3): 10-20.

Sachs, Aaron. 1995. *Eco-Justice: Linking Human Rights and the Environment*.

Worldwatch Paper 127. Washington, DC: Worldwatch Institute.
Sachs, Aaron. 1994. "Men, Sex, and Parenthood in an Overpopulated World." *World Watch Magazine* 7(2).
Wenner, Lambert N. 1992. *Minerals, People, and Dollars: Social, Economic, and Technological Aspects of Mineral Resource Development*. Missoula, MT: USDA Forest Service, Northern Region.

Chapter 7: Agriculture, Nutrition, and Health Issues

Endnotes:
1. World Resources Institute. 1990. *World Resources*. New York: Oxford University Press. Pp. 87-88.
2. Brown, Lester, and Hal Kane. 1994. *Full House: Reassessing the Earth's Carrying Capacity*. New York: W.W. Norton & Co. Pp. 22-24.
3. Commoner, Barry. 1990. *Making Peace with the Planet*. New York: Pantheon Books. Pp. 85-86.
4. Durning, Alan B. and Holly B. Brough. 1991. *Taking Stock: Animal Farming and the Environment*. Worldwatch Paper 103. Washington, DC: Worldwatch Institute.
5. *Statistical Abstract of the United States*. 1995.
6. World Resources Institute. 1990. *World Resources, 1990-1991*. Pp. 56-57.
7. Platt, Anne E. 1994. "Why Don't We Stop Tuberculosis?" *World Watch*. 7(4): 31-34.
8. Garrett, Laurie. 1995. *The Coming Plague: Newly Emerging Diseases in a World Out of Balance*. New York: Farrar, Straus, and Giroux. P. 561.
9. Garrett. *Ibid*. P. 438.
10. Colborn, Theo, John Myers, and Dianne Dumanski. "Hormonal Sabotage." *Natural History* 105 (3): 42-49.
11. Dold, Catherine. 1996. "Hormone Hell." *Discover* 17(9): 52-59.
12. Chivian, Eric. 1994. "The Ultimate Preventive Medicine." *Technology Review* 97(8). Pp. 34-40.
13. Baumgardt, Bill R. and Marshall A. Martin, Coed. 1991. *Agricultural Biotechnology: Issues and Choices*. West Lafayette, IN: Purdue University Agricultural Experiment Station. Pp. 3-21.
14. Rhoades, Robert E. 1991. "The World's Food Supply at Risk." *National Geographic*. 179(4): 74-105.
15. Baumgardt. 1991. Cited above. Pp. 155-156.
16. Baumgardt. 1991. Cited above. Pp. 81-93.

Pertinent Readings:
The Amicus Journal. 1993. Special Section: Biotechnology & Ecology. 15(1).
Barbier, Edward B. 1989. "Sustaining Agriculture on Marginal Land." *Environment* 31(9): 12-17, 36-40.

Brown, Lester R. 1995. "Facing Food Scarcity." *World Watch* 8(6). 10-20.
Brown, Lester. 1995. *Who Will Feed China: Wake-up Call for a Small Planet.* New York: W.W. Norton & Co.
Dunlap, Riley E., and Angela G. Mertig. 1992. *American Environmentalism: The U.S. Environmental Movement,* 1970-1990. Bristol, PA: Crane Russak.
Franklin, Deborah. 1995. "The Poisoning at Pamlico Sound." *Health* 9(5): 108-116.
Fumento, Michael. 1993. *Science Under Siege: Balancing Technology and the Environment.* New York: William Morrow & Co.
Nielsen, Lewis T. 1991. "Mosquitoes Unlimited." *Natural History* (July, 1991). Lead article in a special issue on mosquitoes and the diseases they transmit.
Norse, David. 1992. "A New Strategy for Feeding a Crowded Planet." *Environment* 34(5): 6-11ff.
Platt, Anne. 1996. *Infecting Outselves: How Environmental and Social Disruptions Trigger Disease.* Worldwatch Paper 129. Washington, DC: Worldwatch Institute.
Platt, Anne. 1995. "The Resurgence of Infectious Diseases." *World Watch* 8(4): 26-32.
Rhoades, Robert E. 1991. "The World's Food Supply at Risk." *National Geographic.* 179(4): 74-105.
Time. 1996. Special fall issue: "The Frontiers of Medicine."

Chapter 8: The Environmental Movement

Endnotes:
1. Cahn, Robert and Patricia. 1990. "Did Earth Day Change the World?" *Environment* 32(7): 17-20, 36-42.
2. Louis Harris and Associates. 1988. *Public and Leadership Attitudes to the Environment in Four Continents: A Report of a Survey in Fourteen Countries.* New York: Louis Harris and Associates. Also: *American Results of the UNEP Survey.*
3. Lewis, Martin W. 1992. *Green Delusions.* Durham, NC: Duke University Press
4. Sitarz, Daniel, Ed. 1993. *Agenda 21: The Earth Summit Strategy to Save Our Planet.* Boulder, CO: Earth Press.

Pertinent Readings:
Brundtland, Gro Harlem. 1989. "Our Common Future." *Environment* 31(5): 16-20, 40-43.
Carson, Rachel. 1962. *Silent Spring.* Greenwich, CN: Fawcett Crest.
Commoner, Barry. 1972. *The Closing Circle: Nature, Man, and Technology.* New York: Alfred Knopf. Also: 1990. *Making Peace with the Planet.* New York: Pantheon Books.
Dunlap, Riley E. and Angela G. Mertig. 1992. *American Environmentalism: The U.S. Environmental Movement, 1970-1990.* Bristol, PA: Crane Russak.
Ehrlich, Paul R. 1968. *The Population Bomb.* New York: Ballantine Books.

French, Hillary. G. 1995. "Forging a New Global Partnership." *State of the World*. New York: W.W. Norton & Co.
Lewis, Martin. 1992. *Green Delusions*. Durham, NC: Duke University Press.
Leopold, Aldo. 1949. *A Sand County Almanac*. New York: Oxford Press.
Meadows, Dennis L., et al. 1972. *The Limits to Growth*. New York: New American Library.
Osborn, Fairfield. 1948. *Our Plundered Planet*. New York: Pyramid Books.
Roth, Dennis M. 1988. *The Wilderness Movement and the National Forests*. College Station, TX: Intaglio Press.
Shabecoff, Philip. 1993. *A Fierce Green Fire: The American Environmental Movement*. New York: Hill and Wang.
Schumacher, E.F. 1973. *Small Is Beautiful: Economics As If People Mattered*. New York: Harper and Row.
Tunali, Odil. 1996. "Habitat II: Not Just Another 'Doomed Global Conference.'" *World Watch* 9(3): 32-34.
Ward, Barbara, and Rene Dubos. 1972. *Only One Earth: The Care and Maintenance of a Small Planet*. New York: W.W. Norton.
Wild, Peter. 1979. *Pioneer Conservationists of Western America*. Missoula, MT: Mountain West.

Chapter 9: Optimism or Despair?

Pertinent Readings:
Bailey, Ronald. 1993. *Ecoscam: The False Prophets of Ecological Apocalypse*. New York: St. Martins Press.
Bailey, Ronald, Ed. 1995. *The True State of the Planet*. New York: The Free Press.
Brown, Lester R. 1991. *Saving the Planet: How to Shape an Environmentally Sustainable Global Economy*. New York: W.W. Norton & Co.
Commoner, Barry. 1990. *Making Peace with the Planet*. New York: Pantheon Books.
Easterbrook, Gregg. 1995. *A Moment on the Earth: The Coming Age of Environmental Optimism*. New York: Viking Penguin, 1995.
Ehrlich, Paul R. and Anne H. Ehrlich. 1990. *The Population Explosion*. New York: Simon and Schuster. Also: 1991. *Healing the Planet*. Reading, MA: Addison-Wesley Publishing Co.
Gore, Al. 1993. *Earth in the Balance: Ecology and the Human Spirit*. New York: Plume.
Gordon, Anita, and David Suzuki. 1991. *It's a Matter of Survival*. Cambridge, MA: Harvard University Press.
Gribbin, John. 1990. *Hothouse Earth: The Greenhouse Effect and Gaia*. New York: Grove Weidenfeld.
Helvarg, David. 1994. *The War Against the Greens*. San Francisco: Sierra Club Books.
Hohm, Charles F., Ed. 1995. *Population: Opposing Viewpoints*. San Diego, CA:

Greenhaven Press.
Lewis, Martin W. 1992. *Green Delusions*. Durham, NC: Duke University Press.
Lovelock, James. 1990. *The Ages of Gaia: A Biography of Our Living Earth*. New York: Bantam Books.
Manes, Christopher. 1990. *Green Rage: Radical Environmentalism and the Unmaking of Civilization*. Boston: Little Brown & Co.
McKibben, Bill. 1989. *The End of Nature*. New York: Random House.
Meadows, Donald and Donella. 1992. *Beyond the Limits: Confronting Global Collapse, Envisioning a Sustainable Future*.
Ramphal, Shridath. 1992. *Our Country, Our Planet*. Washington, DC: Island Press.
Ray, Dixy Lee. 1993. *Environmental Overkill: Whatever Happened to Common Sense?* New York: Harper Collins.
Simon, Julian. *The Ultimate Resource*. Princeton, NJ: Princeton University Press.
Simon, Julian, and Herman Kahn, Eds. 1984. *The Resourceful Earth: A Response to Global 2000*. New York: Blackwell.
U.S. Department of State and the Council on Environmental Quality. 1980. *Global 2000 Report to the President*. New York: Pergamon.
Weiner, Jonathan. 1990. *The Next Hundred Years: Shaping the Fate of our Living Earth*. New York: Bantam Books.
Wildavsky, Aaron. 1995. *But Is It True? A Citizen's Guide to Environmental Health and Safety Issues*. Cambridge, MA: Harvard University Press.
Zey, Michael G. 1994. *Seizing the Future*. New York: Simon and Schuster.

Chapter 10: Politics and Reform

Endnotes:
1. *Christian Science Monitor*. 1994. "Environment Council Survives." Oct. 27, 1944.
2. Chadd, Edward A. 1995. "Manifest Subsidy." *Common Cause Magazine* 21(3): 18-21.
3. "Logging Without Laws." 1996. *Audubon* 98(1): 14-18.
4. Kosova, Weston. "Ways to Skin the Act." *Audubon* 98(1): 40-58.
5. Boyle, Robert H. 1993. "All the News That's Fit to Twist." *The Amicus Journal* 15(3): 9
6. O'Keeffe, Michael and Kevin Dailey. 1993. "Checkering the Right." *Buzzworm: The Environmental Journal* 5(3): 38-44.
7. Lewis, Martin W. 1993. *Green Delusions*. Durham, NC: Duke University Press.
8. Shabecoff, Philip. 1993. *A Fierce Green Fire: The American Environmental Movement*. New York: Hill and Wang.

Pertinent Readings:
Bergman, B.J. 1996. "Newt's Game" *Sierra* 81(5): 30-35.

Bernards, Neal. Ed. 1991. *The Environmental Crisis: Opposing Viewpoints*. San Diego: Greenhaven Press.

Caldwell, Lynton K. 1990. *Between Two Worlds, the Environmental Movement and Policy Choice*. New York: Cambridge University Press.

Diamond, Sara. 1995. *Roads to Dominion: Right-Wing Movements and Political Power in the United States*. New York: Guilford Press.

Durbin, Kathie. 1996. "High Noon in the National Forests" (on biodiversity). *The Amicus Journal* 18(2): 26-30.

Franklin, Ben A., Ed. "Washington's Conservative Think Tanks Influence Government Policy." 1996. *The Washington Spectator* 22(8).

Helvarg, David. *The War Against the Greens*. 1994. San Francisco: Sierra Club Books.

Israelson, David. 1990. *Silent Earth: The Politics of Our Survival*. Toronto, Ontario: Penguin Books.

Motavalli, Jim and Ellen Miller. 1996. "Dirty Money." E: *The Environmental Magazine* 8(5): 28-37.

Shabecoff, Philip. 1996. "Greens Vs. Congress: A Play-by-Play." *The Amicus Journal* 18(3): 24-29.

Chapter 11: Bold Actions and Mixed Results

Endnotes:

1. Levin, Ted. 1996. "The Comeback Connecticut." *Sierra* 81(2): 78-79.

2. 1991. "Wetland Alchemy in Arcata." *Sunset* (Mar 91).

3. Stegner, Will, and Jon Bowermaster. 1991. "Garbage Must Be Reduced at Its Source." *The Environmental Crisis*. Neal Bernards, Ed. San Diego: Greenhaven Press.

4. Adler, Robert. 1994. "The Clean Water Act: Has it Worked?" *EPA Journal* 20(1-2): 10-15.

5. Bowermaster, Jon. 1993. "A Town Called Morrisville." *Audubon* 95(4): 42-51.

6. Watkins, T.H. "What's Wrong with the Endangered Species Act?" *Audubon* 98(1): 36-42.

7. Stolzenburg, William. 1996. "A Plate of Tainted Fish." *Nature Conservancy* 42(2): 6.

8. Batisse, Michael. 1990. "Probing the Future of the Mediterranean." *Environment* 32(5): 5-9, 28-34.

9. *Earth Journal Environmental Almanac and Resource Directory*. 1993. Boulder, CO: Buzzworm, Inc. Pp. 101-103.

10. Easterbrook, Gregg. 1995. *A Moment on Earth*. New York: Penguin Books. Pp. 310-312.

11. Bennett, Graham. 1991. "The History of the Dutch National Environmental Policy Plan." *Environment* 33(7): 6-9, 31-36.

Pertinent Readings:
Brown, Lester R., Christopher Flavin, and Sandra Postel. 1991. *Saving the Planet: How to Shape an Environmentally Sustainable Economy.* New York: W.W. Norton & Co.
Commoner, Barry. 1990. *Making Peace with the Planet.* New York: Pantheon Books.
Dopyera, Caroline. 1996. "The Iron Hypothesis" (to reduce global warming). *Earth: The Science of Our Planet* 5(5): 26-33. *EPA Journal* 20(1-2). 1994. Special Issue: "Clean Water Agenda."
Flanagan, Ruth. 1996. "Engineering a Cooler Planet." *Earth: The Science of Our Planet* 5(5): 34-39.
Global Ecology Coalition. 1990. *The Global Ecology Handbook: What You Can Do About the Environmental Crisis.* Boston: Beacon Press.
Lenssen, Nicholas. 1992. *Empowering Development: The New Energy Equation.* Worldwatch Paper 120. Washington, DC: Worldwatch Institute.
Luoma, Jon R. 1996. "Biography of a Lake." *Audubon* 98(5): 66-72.
Naar, John. 1990. Design for a Livable Planet: *You Can Help Clean Up the Environment.* New York: Harper and Row.
Postel, Sandra. 1992. Last Oasis: Facing Water Scarcity. New York: W.W. Norton & Co.
Rauber, Paul. 1996. "An End to Evolution." *Sierra* 81(1): 28-33, 123.
Steger, Will, and Jon Bowermaster. 1990. *Saving the Earth: A Citizen's Guide to Environmental Action.* New York: Alfred A Knopf.
Young, John E. 1993. "Global Network: Computers in a Sustainable Society." Worldwatch Paper 115. Washington, DC: Worldwatch Institute.
Young, John E., and Aaron Sachs. 1994. *The Next Efficiency Revolution: Creating a Sustainable Materials Economy.* Washington, DC: Worldwatch Institute.

Chapter 12: Some Conclusions and a Look Ahead

Pertinent Readings:
Brown, Lester R., and Hal Kane. 1994. *Full House: Reassessing the Earth's Population Carrying Capacity.* New York: W.W. Norton & Co.
Gore, Al. 1993. *Earth in the Balance: Ecology and the Human Spirit.* New York: Penguin-Plume.
Hardin, Garrett. 1986. *Filters Against Folly: How to Survive Despite Economists, Ecologists, and the Merely Eloquent.* New York: Penguin.

Index

Person Index

Note: for additional citations, see Endnotes and Pertinent Readings.

Abbey, Edward, 120

Audubon, John James, 113

Bailey, Ronald, 130-131

Borlaug, Norman, 100

Boulding, Kenneth, 41

Bowermaster, Jon, 162

Brown, Lester R., 128

Brundtland, Gro Harlem, 135

Bush, George, 146-147, 165,

Carson, Rachel, 115-116, 125-126, 173

Carter, Jimmy, 117, 146

Commoner, Barry, 101, 116, 118, 135

Clinton, Bill, 147

Easterbrook, Gregg, 134-135

Cousteau, Jacque, 115

Dubos, Rene, 117

Ehrlich, Paul and Anne, 116, 126

Eisenhower, Dwight D., 121

Ford, Gerald, 146

Forrester, Jay, 126

French, Hilary F., 189

Garrett, Laurie, 106-107,

Gordon, Anita, 128

Gore, Al, 133-134, 147

Hardin, Garrett, 116

Jefferson, Thomas, 113

Johnson, Lyndon, 145, 160

Kahn, Herman, 117

Leopold, Aldo, 114-115

Lewis, Martin, 120

Lovelock, James, 129

Malthus, Thomas, 16, 113

Margulis, Lynn, 129

Marshall, Robert. 115

McKibbin, Bill, 129

Meadows, Dennis and Donella, 116, 126, 128, 174

Muir, John, 114

Nixon, Richard M., 121, 146, 160

Osborn, Fairfield, 115, 173

Pinchot, Gifford, 114

Ray, Dixy Lee, 131

Ramphal, Shridath, 135-136

Reagan, Ronald, 133, 146

Rockefeller, Nelson, 121

Roosevelt, Theodore, 114

Ruckelshaus, William, 146

Simon, Julian, 117, 126, 132

Sitarz, Daniel, 189

Steinbeck, John, 64

Suzuki, David, 128

Thoreau, Henry, 113

Vogt, William, 115

Ward, Barbara, 117

Wattenberg, Benjamin, 132

Wildavsky, Aaron, 134

Wilson, Edward, 74

Zey, Michael, 132

Subject Index

Africa

  population trends, 20, 181-183

  poverty in, 101-102

agriculture

  Green Revolution, 99-101, 127

  intensive, 6

  revolution in technology, 8, 44, 108-110

  trends in production, 99-103

  See also animals, croplands

AIDs, 106

air pollution, 67-69

animals, domestic, 50

house cats killing birds, 75

livestock vs. grains, 103

Asia, population trends, 20-21, 181-183

atmosphere, composition of, 67-69

Audubon Society, 115, 119

  See also environmental organizations

automation, 9

automobile

  changes due to, 80

  number of, 32, 34, 36

  pollution from, 68-69

biotechnology, agricultural, 108-110

birds, extinction of, 75

business cycles, 94-95

Canada, population trends, 19, 181

carrying capacity, global, 3-4, 175

China

  expanding economy, 29

  food production, 102

  population growth, xxvi, 13

population policy, 176

Clean Air Act, 160-161

Clean Water Act, 162

clearcutting, 66

   See also forests, timber

Club of Rome, 126

coal, 44-47, 68-69

colonialism, 87, 88

commercial revolution, 7

Congress, U.S., xvii, 145-148

   See also environmental legislation

Connecticut River, 157

conservative view, 117, 121, 143-144

consumer class, 29

consumption of goods and services

   cheaper goods, 31

   developing nations, 31-32, 36-37

   energy as a measure of, 33-34

   increasing per capita

   industry's role, 37-39

nonmarket consumption

  U.S. example, xxvi-xxvii, 1, 32-36

cities. See urbanization

Council on Environmental Quality, 117, 122, 146-47, 180

Council on Sustainable Development, 147

croplands, 51, 102-104

  converting forests to, 104

cultural resources, 54

deforestation, 65-67

  See also forests, timber

demography, 16-17, 21

depletion of natural resources, 41-55

  U.S. experience, 61-62

  See also individual resources, e.g., forests, fisheries

desertification, 65

developed nations, 10, 30-32

  See also industrial nations

developing nations, 10-11, 27, 31-32

  consumption in, 31-32, 36-37

  prospects for, 135-136, 176-177

development, external costs, 42, 61-62, 83-84, 133-134

disease epidemics, 23, 105-107

Dutch National Environmental Plan, 166

Earth Day, xvii, xxiii, 116, 119, 125

Earth First!, 120, 151

ecology, global

  components, xxviii

  ecological perspective, 59-61

  Gaia concept, 127

  human factor critical in, xxviii-xxix, 139-140

economic efficiency, 2-4, 81-82

economic growth, 7-8, 79-86

  benefits and costs of, 81-84, 125-135

  external costs, 42

economic systems, types of, 4-10

education, 26-27, 151-152, 180

  See also Earth Day

Electronic Age, 9-10

endangered species, 122, 163-164

  See also extinction

energy

    conservation, xviii, 154-155

    consumption, 33-34, 154-156

    Industrial Revolution and, xxiv, 8

    sources, 44-47, 154-156

environment

    components of, xxviii-xxix, 42

    degradation of, 59-78, 111-115

    effects of changes, 108

    resiliency after abuse, 77-78

    See also specific resources, e.g., forests, oceans, etc.

environmental dilemma, global dimensions, xxiii-xxix, 41-42

    basis for concern, optimism, or both, xxx-xxxi, 125-135

    See also reform of environmental policies

environmental impacts, 61-77

    as basis for environmental movement, 115-118

    monitoring trends, 80-81

    response options, 128, 155-160, 172-180

    See also specific problems, resources

environmental justice, 94

environmental movement, 113-118, 127-129, 168-169,

environmental organizations, 119, 151

environmental protection

   barriers to reform, 139-143, 169-170

   laws and agreements, 121-123, 160-166, 184-189

   results of legislation, 160-166, 173-174

   U.S. experience, 145-148

Environmental Protection Agency, 145-147, 162-163

epidemics, 23, 105-107

ethnocentrism, 89-80

extinction of species, 50, 74-76

extremism, environmental

   common elements, 149-150, 169

   radical, 120, 148-151

   reactionary, 121, 148-151

   religious, 130, 149-150

family planning, 26-27, 104, 159-160, 172-173

famine, 23, 99, 111-112

fisheries, 53, 73-75

food chain, 60, 190n1

forests, 47-49, 65-67

  tropical rain forests, 47, 66-67

  See also timber

fossil fuels, 8, 44-47

  alternatives to, 154-156

  pollution from, 67-79

free-trade agreements, 12

Germany, 87-88

Global Fishing Treaty, 165

global marketplace, 11, 79-80

Global 2000 Report, 117, 146

global warming, 69-70

  Global Warming Treaty, 165

Great Depression, 95

Great Migration, 25, 72

"green" enterprises, 86-87

Green Revolution, 100-101, 127

gross domestic product (GDP) of nations, 30-32

herding economy, 6

horticultural economy, 5

human factor in

   causing environmental impacts, xxviii-xxix, 61-62, 112-113

   resolving impacts, xxix-xxxi, 172-174, 180

human origins, 111-112

Humboldt Bay, 157

hunger, 98-99

hunting and gathering economy, 5

ice ages, 1, 111

immigration

   illegal, 92-93

   legal, 25, 173

India, 15, 72

industrial nations, 10, 19-20

   as consumers, 30-36, 41-42

Industrial Revolution, xxiv-xxv, 7-8, 79-80, 112, 167-168

industry

   modernization of, 7-10

   pro-development philosophy, 130-133

   role in encouraging consumption, 37-39

   role in politics, 147-148

Lake Erie, 158

Latin America, 21, 102, 181-182

liberal view on the environment, 144-145

limits of growth, 126

litter, 73

livestock, 60, 103

   also see animals, domestic

Louisiana Purchase, 113

malnutrition, 99, 105

minerals

   metals and industrial minerals, 43-44

   also see fossil fuels

migration, 17, 25, 92, 173

mining and drilling, 63-64

Montreal Protocol, 164, 185

National Environmental Policy Act, 121-122, 147, 161-162

natural gas. See petroleum

noise and eye pollution, 73

nuclear power, 8, 155

ocean dumping ban, 164, 184-185

oil. See petroleum

ozone depletion, 70-71, 185

petroleum,

   drilling for, 63-64

   natural gas, 46-47, 156

   oil, 44-47

   pollution from, 67-70

pesticides, 71-72, 107-108

photosynthesis, 190n1

plants becoming extinct, 50, 70-73

plastics, 55

poaching, 65

polls. See public opinion

pollution, 67-73

   See also specific type; e.g., air, water, soil, etc.

population, 21

   consequences of growth, 15, 41-42, 116, 127-129

   economic systems and, 2-10

   in developing nations, 27

   rapid growth explained, 22-25, 104

slow growth, 26

technology and, 2-3, 23-24

trends, xxvi, 13-15, 22, 172-173

post-industrial era, 9-10

poverty, 93-94, 98, 101-102

private property ethic, 113-114, 141

profit as motive for industry, 81-82

public opinion polls on

  environmental protection, xviii, 119

  immigration, 93

recreation, outdoor, 54-55

recycling, 158-159, 180

reform of environmental policies

  approaches to, 133-135, 121-123, 173-174, 179-181

  costs of, 177-178

  emerging consensus on, 136, 170-172

  international cooperation on, 122-123, 174, 184-189

  obstacles to, 139-143, 169-170

  success of policies, 159-166

religion, role of, 130

in family planning, 20, 159-160

in politics, 11, 130, 140-141, 149-150

resources, renewable and nonrenewable, 42-54

research, environmental, 125, 140, 144, 168-170

Rio Conference, 119, 185-187

sanitation and health, 23

social problems, 89-95, 105-107

social status, wealth as a measure of, 39-40, 91

socioeconomic systems, 4-10

soil depletion, 64-65, 102-103

Spaceship Earth concept, 41

special interests, 119-121, 136-137, 146, 149-151, 170

standard of living, 29-40, 175-176

  defined, 33

  extremes in, 29-32, 35-36

  redefined, 39-40

  U.S. example, 32-36

synthetic materials, 55

technology, 1-12

  early development of, xxiv, xxvii, 2-10

environment and, 10-11, 41-42, 61-62 111-113, 167-168

increasing economic efficiency, 3-4

potential of, 12, 109, 118, 130-131, 137, 174-175

stimulates change, 2-4, 84-86

theory of demographic transition, 26, 104

think tanks, 144

Third World nations, 191n5

See also developing nations

timber harvesting, 48-49, 66-67

salvage rider, 148

unsustainable yields, 48, 65-67

See also forests

toxic chemicals, 73, 107-108, 162-163

transportation and communication, 9-10, 156

tropical forests. See forests

Union of Concerned Scientists, 151-152

United Nations

Environmental Programme, 56-57

international conferences, 117, 119, 122-123, 184-189

report on climate change, 165

United States

consumption patterns, 32-36

leadership needed, 178-179

population trends, 19, 25, 181-182

U.S. Forest Service, 114-115, 146

urbanization, 8, 13-15, 85-86, 181-183

values, cultural, 39-40, 141-142

wars and executions, 24-25

waste disposal, xxix, 76-77, 158-159

water, fresh, 35, 51-53

water pollution, 71-72, 101

wetlands, 49

wilderness, 121, 161

Wilderness Society, 115, 119

wildlife extinction, 50, 70-73

women, status of, 26, 89, 173, 188

World Bank, 57, 166

World Resources Institute, 56-57, 117

Biographical Sketch

Lambert N. "Bert" Wenner is a Syracuse University graduate (PhD, 1970) and taught sociology, anthropology, and political science at the University of Idaho, in the University Maryland's European and Asian Divisions, and at other institutions. More recently he was employed by the U.S. Forest Service in successive roles as regional sociologist, environmental specialist, and national coordinator for social impact analysis. Wenner also served as Assistant Director for Environmental Quality in the USDA Office of Agricultural Biotechnology in Washington, DC. Wenner is the author of several articles and book chapters, and numerous agency publications on social and environmental topics. He lived 10 years abroad and has traveled extensively in Europe, North America, and the Far East. Now retired to Salem, Oregon, Wenner works part time as a social and environmental impacts consultant.